Nanogenerators in Korea

Nanogenerators in Korea

Special Issue Editors

Dukhyun Choi
Yong Tae Park

MDPI • Basel • Beijing • Wuhan • Barcelona • Belgrade

Special Issue Editors
Dukhyun Choi Yong Tae Park
Kyung Hee University Myongji University
Korea Korea

Editorial Office
MDPI
St. Alban-Anlage 66
4052 Basel, Switzerland

This is a reprint of articles from the Special Issue published online in the open access journal *Micromachines* (ISSN 2072-666X) from 2018 to 2019 (available at: https://www.mdpi.com/journal/micromachines/special_issues/nanogenerators)

For citation purposes, cite each article independently as indicated on the article page online and as indicated below:

LastName, A.A.; LastName, B.B.; LastName, C.C. Article Title. *Journal Name* **Year**, *Article Number*, Page Range.

ISBN 978-3-03897-622-6 (Pbk)
ISBN 978-3-03897-623-3 (PDF)

© 2019 by the authors. Articles in this book are Open Access and distributed under the Creative Commons Attribution (CC BY) license, which allows users to download, copy and build upon published articles, as long as the author and publisher are properly credited, which ensures maximum dissemination and a wider impact of our publications.

The book as a whole is distributed by MDPI under the terms and conditions of the Creative Commons license CC BY-NC-ND.

Contents

About the Special Issue Editors vii

Yong Tae Park and Dukhyun Choi
Editorial for the Special Issue on Nanogenerators in Korea
Reprinted from: *Micromachines* **2019**, *10*, 97, doi:10.3390/mi10020097 1

Dahoon Ahn and Kyungwho Choi
Performance Evaluation of Thermoelectric Energy Harvesting System on Operating Rolling Stock
Reprinted from: *Micromachines* **2018**, *9*, 359, doi:10.3390/mi9070359 3

Hyun-Woo Park, Nghia Dinh Huynh, Wook Kim, Hee Jae Hwang, Hyunmin Hong, KyuHyeon Choi, Aeran Song, Kwun-Bum Chung and Dukhyun Choi
Effects of Embedded TiO_{2-x} Nanoparticles on Triboelectric Nanogenerator Performance
Reprinted from: *Micromachines* **2018**, *9*, 407, doi:10.3390/mi9080407 15

Youn-Hwan Shin, Inki Jung, Hyunchul Park, Jung Joon Pyeon, Jeong Gon Son, Chong Min Koo, Sangtae Kim and Chong-Yun Kang
Mechanical Fatigue Resistance of Piezoelectric PVDF Polymers
Reprinted from: *Micromachines* **2018**, *9*, 503, doi:10.3390/mi9100503 25

Jin Pyo Lee, Jae Won Lee and Jeong Min Baik
The Progress of PVDF as a Functional Material for Triboelectric Nanogenerators and Self-Powered Sensors
Reprinted from: *Micromachines* **2018**, *9*, 532, doi:10.3390/mi9100532 33

Moonwoo La, Jun Hyuk Choi, Jeong-Young Choi, Taek Yong Hwang, Jeongjin Kang and Dongwhi Choi
Development of the Triboelectric Nanogenerator Using a Metal-to-Metal Imprinting Process for Improved Electrical Output
Reprinted from: *Micromachines* **2018**, *9*, 551, doi:10.3390/mi9110551 46

Jihoon Chung, Deokjae Heo, Banseok Kim and Sangmin Lee
Superhydrophobic Water-Solid Contact Triboelectric Generator by Simple Spray-On Fabrication Method
Reprinted from: *Micromachines* **2018**, *9*, 593, doi:10.3390/mi9110593 55

Kwangseok Lee, Jeong-won Lee, Kihwan Kim, Donghyeon Yoo, Dong Sung Kim, Woonbong Hwang, Insang Song and Jae-Yoon Sim
A Spherical Hybrid Triboelectric Nanogenerator for Enhanced Water Wave Energy Harvesting
Reprinted from: *Micromachines* **2018**, *9*, 598, doi:10.3390/mi9110598 64

Wonjun Jang, Hyun A Cho, Kyungwho Choi and Yong Tae Park
Manipulation of *p*-/*n*-Type Thermoelectric Thin Films through a Layer-by-Layer Assembled Carbonaceous Multilayer Structure
Reprinted from: *Micromachines* **2018**, *9*, 628, doi:10.3390/mi9120628 75

Mario Culebras, Kyungwho Choi and Chungyeon Cho
Recent Progress in Flexible Organic Thermoelectrics
Reprinted from: *Micromachines* **2018**, *9*, 638, doi:10.3390/mi9120638 85

Hee Jae Hwang, Younghoon Lee, Choongyeop Lee, Youngsuk Nam, Jinhyoung Park, Dukhyun Choi and Dongseob Kim
Mesoporous Highly-Deformable Composite Polymer for a Gapless Triboelectric Nanogenerator via a One-Step Metal Oxidation Process
Reprinted from: *Micromachines* **2018**, *9*, 656, doi:10.3390/mi9120656 . . . . **121**

Wonwoo Lee, Yonghee Jung, Hyunseung Jung, Chulhun Seo, Hosung Choo and Hojin Lee
Wireless-Powered Chemical Sensor by 2.4 GHz Wi-Fi Energy-Harvesting Metamaterial
Reprinted from: *Micromachines* **2019**, *10*, 12, doi:10.3390/mi10010012 . . . . **132**

Minki Kang, Tae Yun Kim, Wanchul Seung, Jae-Hee Han and Sang-Woo Kim
Cylindrical Free-Standing Mode Triboelectric Generator for Suspension System in Vehicle
Reprinted from: *Micromachines* **2019**, *10*, 17, doi:10.3390/mi10010017 . . . . **141**

About the Special Issue Editors

Dukhyun Choi, Dr., received his Ph.D. in Mechanical Engineering from Pohang University of Science and Technology (Postech) in 2006. From 2006–2008, he was a postdoctoral fellow with Dr. Luke P. Lee at UC Berkeley, where he studied about nanplasmonics. Dr. Choi moved to Samsung Advanced Institute of Technology (SAIT) as a research staff, where he started to study for energy harvesters. In 2010, he joined in Department of Mechanical Engineering at Kyung Hee University as an assistant professor, and now he is an associate professor. His research interests include Energy Harvesters, Plasmonics, Hybrid Photovoltaics, Flexible Electronics, and Water Splitting. Details can be found at: http://dchoi.khu.ac.kr.

Yong Tae Park, Dr., joined Myongji University as an assistant professor of Mechanical Engineering in March 2014, after his postdoctoral research for 2.5 years in Chemical Engineering and Materials Science at University of Minnesota (PI: Prof. Christopher Macosko). He finished his Ph.D. in Mechanical Engineering at Texas A&M University, with Materials Science emphasis (Advisor: Prof. Jaime Grunlan). He also received B.S. and M.S. in Mechanical Engineering, with Applied Mechanics emphasis, from KAIST and POSTECH, respectively. He spent four years for failure analysis of memory packages and modules at SK Hynix Semiconductor Package R&D Center. His current research interests lie in both the development of multifunctional thin coatings using layer-by-layer technique and the study of mechanically enhanced, electrically conductive, and gas impermeable polymer nanocomposites.

Editorial

Editorial for the Special Issue on Nanogenerators in Korea

Yong Tae Park [1,*] and Dukhyun Choi [2,*]

1. Department of Mechanical Engineering, Myongji University, Yongin, Gyeonggi 17058, Korea
2. Department of Mechanical Engineering, Kyung Hee University, Yongin, Gyeonggi 17104, Korea
* Correspondence: ytpark@mju.ac.kr (Y.T.P.); dchoi@khu.ac.kr (D.C.); Tel.: +82-31-330-6343 (Y.T.P.); +82-31-201-3320 (D.C.)

Received: 28 January 2019; Accepted: 29 January 2019; Published: 29 January 2019

Nanogenerator-based technologies have found outstanding accomplishments in energy harvesting applications over the past two decades. These new power production systems include thermoelectric, piezoelectric, and triboelectric nanogenerators, which have great advantages such as eco-friendly low-cost materials, simple fabrication methods, and operability with various input sources. Since their introduction, many novel designs and applications of nanogenerators as power suppliers and physical sensors have been demonstrated based on their unique advantages. This Special Issue in Micromachines, titled "Nanogenerators in Korea", compiles some of the recent research accomplishments in the field of nanogenerators for energy harvesting. It consists of 12 papers, which cover both the fundamentals and applications of nanogenerators, including two review papers. These papers can be categorized into four groups as follows:

(1) Triboelectric Nanogenerators (TENG). Lee et al. [1] provided an educational review of PVDF-based triboelectric energy harvesters and self-powered sensors. PVDF is a promising dielectric material for energy harvesting due to its interesting multi-faceted properties, which can be further improved through composites. Kang et al. [2] studied energy harvesting from suspension systems of vehicles. Such an energy harvester could support the ADAS technology in autonomous vehicles. Hwang et al. [3] investigated a gapless structure triboelectric nanogenerator using a mesoporous and deformable Al_2O_3–PDMS composite. They also studied its pressure sensitivity and showed its application in smart cushions for monitoring human sitting positions. Lee et al. [4] proposed a spherical TENG structure that utilized both solid–solid contact and liquid–solid contact for water wave energy harvesting. The innovative hybrid design could scavenge greater amounts of energy than the individual methods used separately. Chung et al. [5] investigated an easy-to-fabricate water–solid contact TENG, the surface of which was made superhydrophobic by a simple spray-on technique. The electrical output could be maximized by maintaining a Cassie–Baxter state between the water and the superhydrophobic surface. La et al. [6] proposed a metal-to-metal imprinting process to create micro- and nano-scale structures on the surface of aluminum, which formed one of the layers of the TENG. The nano-structured aluminum showed enhanced output compared to non-structured aluminum. Park et al. [7] investigated the effect of embedding highly dielectric TiO_2 nanoparticles in PDMS to improve the TENG performance. They also demonstrated the output enhancement using a windmill-integrated TENG system.

(2) Thermoelectric Nanogenerators. Culebras et al. [8] provided a comprehensive review of organic thermoelectric materials and their corresponding composites, with a focus on polymers and carbon nanofillers. Strategies to enhance the thermoelectric performance, polymer composite-based thermoelectric devices, and brief conclusions and outlooks for future research were summarized. Ahn et al. [9] designed an optimized thermoelectric energy harvesting system

and applied it on a rolling stock as low-power sensor nodes in a self-powered independent monitoring system. Jang et al. [10] investigated the thermoelectric performance of carbonaceous nanomaterials-based polymeric multilayer structures, showing p-type or n-type thermoelectric properties by simply changing the electrolyte.

(3) Piezoelectric Nanogenerators. Shin et al. [11] investigated the fatigue resistance of piezoelectric PVDF by subjecting the device to 10^7 cycles of tension and compression. The tension experiments showed stable polarization, while the compression experiments showed a 7% decrease in polarization. However, no notable decrease in output voltage was observed.

(4) Metamaterial Nanogenerators. Lee et al. [12] investigated energy-harvesting metamaterials for a novel wireless-powered chemical sensing system. The resonance frequency and voltage output from the metamaterial changed depending on the chemical compound and its concentration in the channel.

We would like to thank all the authors for their papers submitted to this Special Issue. We would also like to acknowledge all the reviewers for their careful and timely reviews to help improve the quality of this Special Issue.

Conflicts of Interest: The authors declare no conflict of interest.

References

1. Lee, J.P.; Lee, J.W.; Baik, J.M. The Progress of PVDF as a Functional Material for Triboelectric Nanogenerators and Self-Powered Sensors. *Micromachines* **2018**, *9*, 532. [CrossRef] [PubMed]
2. Kang, M.; Kim, T.Y.; Seung, W.; Han, J.-H.; Kim, S.-W. Cylindrical Free-Standing Mode Triboelectric Generator for Suspension System in Vehicle. *Micromachines* **2019**, *10*, 17. [CrossRef] [PubMed]
3. Hwang, H.J.; Lee, Y.; Lee, C.; Nam, Y.; Park, J.; Choi, D.; Kim, D. Mesoporous Highly-Deformable Composite Polymer for a Gapless Triboelectric Nanogenerator via a One-Step Metal Oxidation Process. *Micromachines* **2018**, *9*, 656. [CrossRef] [PubMed]
4. Lee, K.; Lee, J.-W.; Kim, K.; Yoo, D.; Kim, D.S.; Hwang, W.; Song, I.; Sim, J.-Y. A Spherical Hybrid Triboelectric Nanogenerator for Enhanced Water Wave Energy Harvesting. *Micromachines* **2018**, *9*, 598. [CrossRef] [PubMed]
5. Chung, J.; Heo, D.; Kim, B.; Lee, S. Superhydrophobic Water-Solid Contact Triboelectric Generator by Simple Spray-On Fabrication Method. *Micromachines* **2018**, *9*, 593. [CrossRef] [PubMed]
6. La, M.; Choi, J.H.; Choi, J.-Y.; Hwang, T.Y.; Kang, J.; Choi, D. Development of the Triboelectric Nanogenerator Using a Metal-to-Metal Imprinting Process for Improved Electrical Output. *Micromachines* **2018**, *9*, 551. [CrossRef]
7. Park, H.-W.; Huynh, N.D.; Kim, W.; Hwang, H.J.; Hong, H.; Choi, K.H.; Song, A.; Chung, K.-B.; Choi, D. Effects of Embedded TiO_{2-x} Nanoparticles on Triboelectric Nanogenerator Performance. *Micromachines* **2018**, *9*, 407. [CrossRef] [PubMed]
8. Culebras, M.; Choi, K.; Cho, C. Recent Progress in Flexible Organic Thermoelectrics. *Micromachines* **2018**, *9*, 638. [CrossRef] [PubMed]
9. Ahn, D.; Choi, K. Performance Evaluation of Thermoelectric Energy Harvesting System on Operating Rolling Stock. *Micromachines* **2018**, *9*, 359. [CrossRef]
10. Jang, W.; Cho, H.A.; Choi, K.; Park, Y.T. Manipulation of p-/n-Type Thermoelectric Thin Films through a Layer-by-Layer Assembled Carbonaceous Multilayer Structure. *Micromachines* **2018**, *9*, 628. [CrossRef]
11. Shin, Y.-H.; Jung, I.; Park, H.; Pyeon, J.J.; Son, J.G.; Koo, C.M.; Kim, S.; Kang, C.-Y. Mechanical Fatigue Resistance of Piezoelectric PVDF Polymers. *Micromachines* **2018**, *9*, 503. [CrossRef]
12. Lee, W.; Jung, Y.; Jung, H.; Seo, C.; Choo, H.; Lee, H. Wireless-Powered Chemical Sensor by 2.4 GHz Wi-Fi Energy-Harvesting Metamaterial. *Micromachines* **2019**, *10*, 12. [CrossRef]

© 2019 by the authors. Licensee MDPI, Basel, Switzerland. This article is an open access article distributed under the terms and conditions of the Creative Commons Attribution (CC BY) license (http://creativecommons.org/licenses/by/4.0/).

Article

Performance Evaluation of Thermoelectric Energy Harvesting System on Operating Rolling Stock

Dahoon Ahn [1] and Kyungwho Choi [2,*]

[1] Advanced Railroad Vehicle Division, Korea Railroad Research Institute, Uiwang-si 16105, Korea; dhahn@krri.re.kr
[2] New Transportation Innovative Research Center, Korea Railroad Research Institute, Uiwang-si 16105, Korea
* Correspondence: kwchoi80@krri.re.kr; Tel.: +82-31-460-5603

Received: 6 July 2018; Accepted: 18 July 2018; Published: 20 July 2018

Abstract: During rolling stock operation, various kinds of energy such as vibration, heat, and train-induced wind are dissipated. The amount of energy dissipation cannot be overlooked when a heavy railroad vehicle operates at high speed. Therefore, if the wasted energy is effectively harvested, it can be used to power components like low power sensor nodes. This study aims to review a method of collecting waste heat, caused by the axle bearing of bogie in a rolling stock. A thermoelectric module (TEM) was used to convert the temperature gradient between the surface of the axle bearing housing and the outdoor air into electric energy. In this study, the output performance by temperature difference in the TEM was lab-tested and maximized by computational fluid analysis of the cooling fins. The optimized thermoelectric energy harvesting system (TEHS) was designed and applied on a rolling stock to analyze the power-generating performance under operation. When the rolling stock was operated for approximately 57 min including an interval of maximum speed of 300 km/h, the maximum open circuit voltage was measured at approximately 0.4 V. Based on this study, the system is expected to be utilized as a self-powered independent monitoring system if applied to a low-power sensor node in the future.

Keywords: energy harvesting; thermoelectric generator; railroad vehicle; rolling stock; axle bearing

1. Introduction

The railway system was a symbol of modernization and has been developed for more than 100 years with its advantages of mass transportation, punctuality, and fast speed. Currently, the advancement of the high-speed train has led to a new era of the railway system as a fast and safe form of public transportation [1–4]. For the safety of passengers within a rapid railway system, not only the reliability of the railway system but also the monitoring technology for understanding abnormal conditions must be improved. The KTX (high-speed railroad system in Korea), for instance, experiences time delays mainly because defects have occurred in the axles and vehicle wheels, which might result in serious problems related to derailments [5,6]. To solve the problem, railway monitoring systems detect and analyze abnormal temperature rises according to axle bearing degradation under operation, with an infrared sensor called the Hot Box Detector (HBD) installed at constant track intervals (approximately 40 km) [7]. Note, however, that errors may occur as fast-moving vehicles are measured with a remotely installed infrared light on the ground; moreover, real-time monitoring is impossible, so it is difficult to respond quickly when an abnormal situation occurs. Therefore, it is necessary to develop an on-board real-time monitoring system for axle bearing inspections. This system must be supplied with power, and this may incur installation costs for electric wires and maintenance costs. As such, harvesting many different kinds of waste energy generated by running vehicles may be the solution to these problems [4,8–32]. This study focused on the axle bearing housing, which is the boundary of high and low temperatures induced by heat conduction from inside the housing due to axle rotation and by cooling from outside due to the forced convection caused by the running vehicle. When a temperature gradient occurs constantly while in operation, waste heat can be collected due to the thermoelectric effect [33–35]. As the direct conversion of electric energy and heat energy, the thermoelectric effect can be classified into two: the Seebeck effect that converts heat energy into electric energy, and the Peltier effect that converts electric energy into heat energy. The Seebeck effect generates power through the difference in electric potential induced by the diffusion movement of the electron carriers. Therefore, installing thermoelectric module (TEM) on the axle bearing housings of boundary surfaces with a temperature difference and maximizing this temperature difference may be effective in generating power. In this study, a modularized thermoelectric energy harvesting system (TEHS) was developed by choosing a suitable TEM through performance test in the lab and optimizing the design of cooling fin in order to keep the given temperature difference as large as possible by utilizing the Seebeck effect. The system was then installed on a Korean high-speed train to analyze the output performance according to the driving condition, and we reviewed whether or not the driving power of the real-time monitoring system can be applied.

2. System Design

A TEHS consists of a TEM and a cooling system. As shown in Figure 1, one side of the TEM is attached to an axle bearing housing as the heat source, generating power by thermoelectric effect. A cooling system attached to the other side of the TEM improves the power-generating amount by increasing the temperature gradient on both sides of the TEM through effective cooling progress. In addition, the cooling system acts to protect the TEM from impact from external substances as the rolling stock runs fast and also serves as an installation frame fixing the system on the housing.

2.1. Thermoelectric Module

Two bulk types of TEM (128A1030, Peltron GmbH., Fürth, Germany and TK-1-3-S, HTRD, Anyang, Republic of Korea) were tested for thermoelectric energy harvesting on the axle bearing housing of the rolling stock. Both TEMs have 20 × 20 mm^2 surface area as the maximum space where the TEM can be installed on the surface of the axle bearing housing. The temperature difference between the axle bearing housing—the high-temperature part while the train vehicle is running—and the outdoor air is approximately 15 °C [4], and the average operation temperature of the TEM is around room temperature; thus, according to this, a generally used commercial TEM optimized at room temperature was selected.

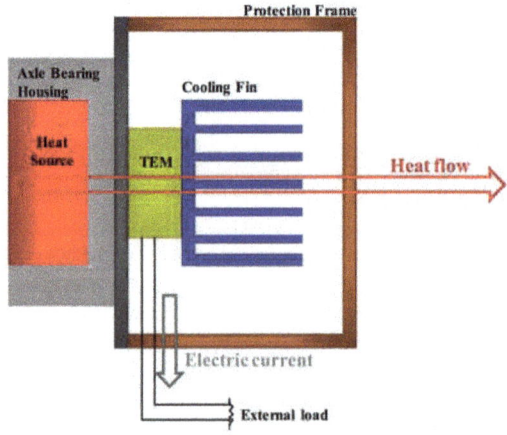

Figure 1. System architecture of thermoelectric energy harvesting system.

2.2. Cooling System

In this study, the heat flux due to the temperature difference between the axle bearing housing of the rolling stock and the outdoor air was converted into electric energy by using a TEM. The thermal energy generated from the axle bearing as the heat source is dependent on the driving conditions of the railroad vehicle. As such, to improve the performance, this study attempted to maximize the air cooling effect by the driving of rolling stocks. Therefore, cooling fins were applied to the TEM, and numerical analysis on heat transfer through the air cooling of the TEM was performed to optimize the shape of the cooling fin. The analysis model is illustrated in Figures 2 and 3. Installed at the bottom part of the TEM are an aluminum plate replicating the axle bearing housing and a size of 40 mm × 40 mm of an electric heater as a heat source.

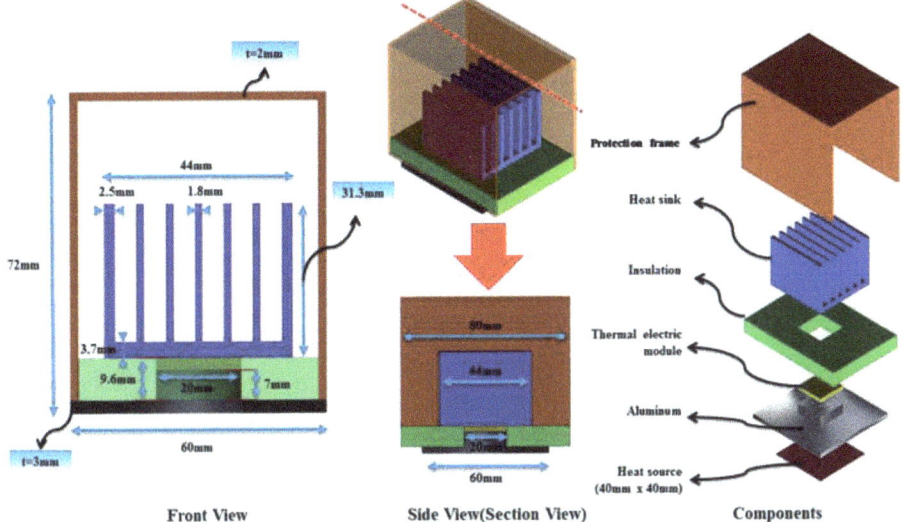

Figure 2. Illustration of the thermoelectric energy harvesting system (TEHS) for application on axle housing of rolling stock.

To prevent unwanted thermal leakage, a thermal insulator is located around the TEM; installed on its upper part are the cooling fins to optimize the air cooling effect. On top of the TEM is a protection frame to protect the cooling fins and the TEM from the impact of ballast flying and substances occurring while driving. In Figure 3, the numerical analysis considers the flow interference and other types of effect by the cooling fins and protection frame; to shorten the calculating time, a bilateral symmetric model and 1-dimensional heat flux were assumed. With the numerical analysis, the temperature gradient of the high- and low-temperature parts of the TEM by changing the number and height of the plate-type cooling fin was analyzed, and the cooling fin shape was optimized to increase the power-generating performance as well.

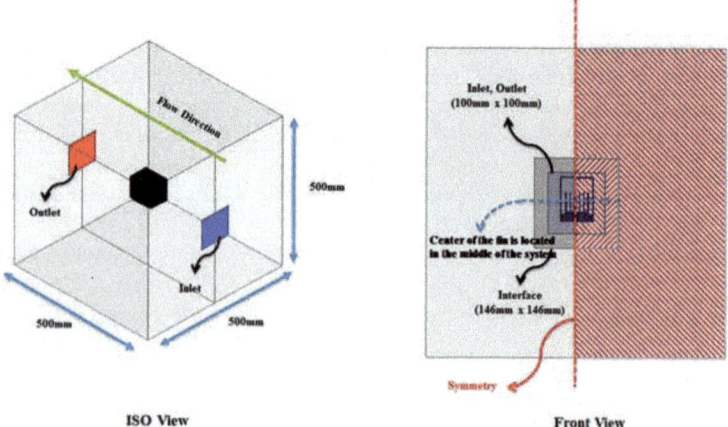

Figure 3. Boundary conditions of fluid region for numerical analysis.

As a result, the speed of the slipstream decreases as the number of cooling fins increases, resulting in flow resistance. Therefore, although the average surface heat transfer coefficient decreases, cooling performance improves by the increase of effective surface contact with air for heat transfer. By reviewing the temperature distribution of the entire system according to the number of cooling fins (Figure 4), the temperature of the low-temperature surface of the heatsink and the TEM decreases, whereas the temperature of the high-temperature surface is constant as the number of cooling fins increases. The configuration of the cooling fin at which its number is over 11 is not considered because there is a limitation regarding manufacturing thin configurations. According to the numerical analysis result by the change of height of the cooling fin (Figure 5), the speed of air passing through the fins decreases due to the interference effect of the fins and protection frame as the height of the fin increases. This may result in a decrease of heat dissipation into the air from the cooling fins; as with the analysis on the numbers of the cooling fin, however, the total heat flux increased due to the increase of the heat transfer area. Thus, a larger temperature gradient was formed on both sides of the TEM. A cooling fin height of more than 51.3 mm was not considered due to the interference of the protection frame. The size of the protection frame was determined by the regulation of vehicle limitation restricting the outermost size considering the attachment to the rolling stock.

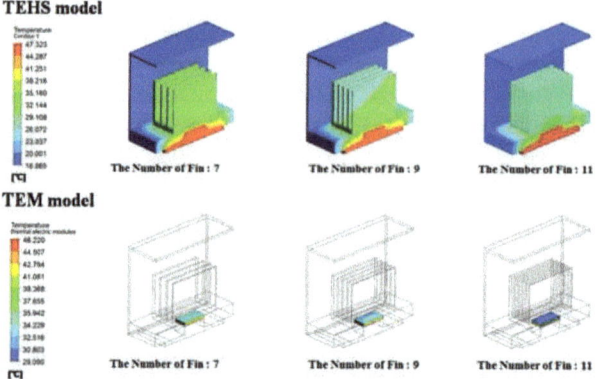

Figure 4. Numerical analysis results of temperature gradient of the TEHS with varying number of cooling fins.

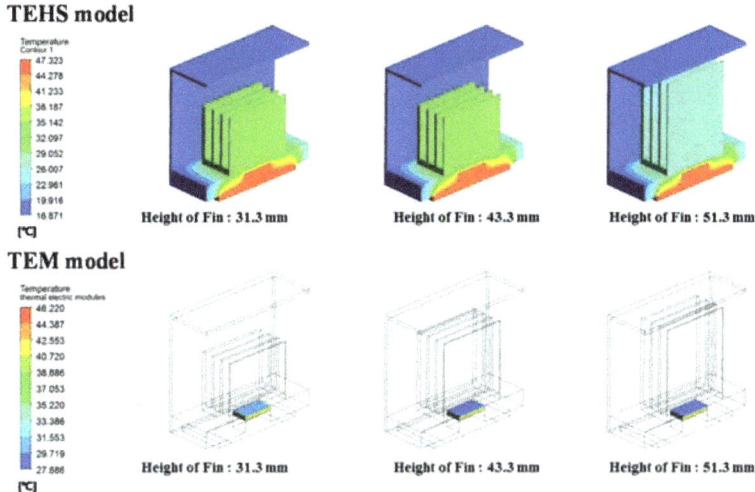

Figure 5. Numerical analysis results of temperature gradient of the TEHS with varying height of cooling fin.

The results of testing by two conditions are shown in Table 1. As the number and height of the fin shape limited by the size of the protection frame increase, the temperature gradient on both sides of the TEM increases. Generally, an air cooling system using cooling fins shows decreasing heat dissipation after reaching the optimized number and height of the fins [36]. Note, however, that the TEHS of this study does not see this kind of result due to the limit of increase in the number and height of the fins. Therefore, a numerical analysis was performed on a bar-type cooling fin that can increase the contact surface area where heat transfer occurs (Figure 6). Bar-type fins with a height of 51.3 mm, which showed optimal results in plate-type cooling fins, were fabricated and arranged at equal intervals in the length and width directions. All the analysis conditions except those mentioned above are the same as the plate-type cooling fins.

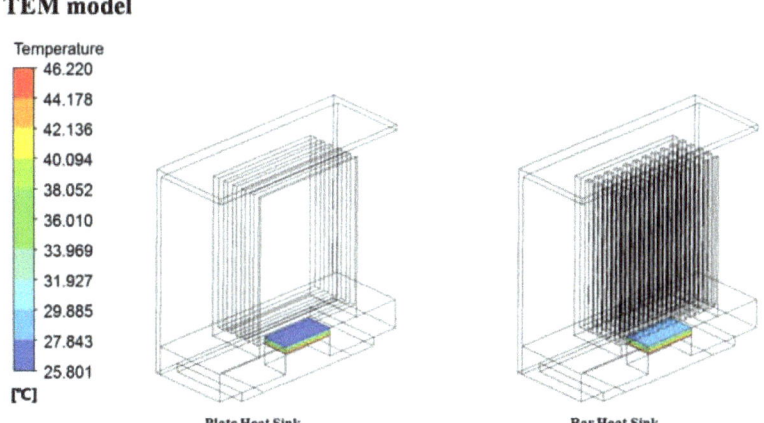

Figure 6. Numerical analysis results of temperature gradient of the thermoelectric module (TEM) with plate and bar type cooling fins.

Table 1. Temperature difference between heating and cooling parts (°C).

Number of Fin	Height of Fin (mm)					
	31.3	35.3	39.3	43.3	47.3	51.3
7	14.89	15.66	16.34	16.94	17.49	17.97
9	16.12	16.93	17.64	18.25	18.81	19.32
11	16.81	17.65	18.39	19.01	19.57	20.05

As a result, the TEHS shows a larger TEM temperate gradient for plate-type cooling fins than bar-type cooling fins. With the characteristics of the shape of the bar-type fins, the bar-type fins have strong turbulent flow between bars resulting, in an increased flow resistance; thus generating high pressure on the inhalation part of the fin. This pressure interrupts the flow inhaled into the system, so the flux flowing in the bar-type cooling fin is 50% of the plate-type cooling fin, and the total heat dissipation decreases. The effective surface on the bar-type cooling fin where the fin comes in contact with air is approximately 42% higher than the plate-type cooling fin, but heat transfer rarely occurs on the back side of the bar-type cooling fin due to the fluid congested between bars. Thus, the available surface area capable of heat transfer somewhat decreases. Based on the analysis result, a cooling system for applying the TEHS on rolling stock was designed.

3. Performance Test on TEMs

A TEHS replicating the condition of running rolling stock was fabricated utilizing a commercial TEM integrated with optimized cooling fins in order to test the output performance of the TEM, as shown in Figure 7. A thermocouple was installed in a 40 mm × 40 mm × 10 mm aluminum plate to measure the temperature of the high-temperature side of the TEM. Attached on one side of the aluminum plate was an additional TEM (HT-15-15-Lq 173, HTRD, Anyang, Korea) as heat source; on its other side, another TEM was installed for power generation. Placed on the upper side of the TEM were cooling fins, with a thermocouple mounted to measure the temperature of the low-temperature side of the TEM. A test model was designed with a cooling fan to simulate the cooling conditions by train-induced wind. The temperature difference between the low- and high-temperature sides of the TEM was controlled within 10–30 °C, similar to the temperature difference between the axle bearing housing and outdoor air when the rolling stock is under operation [4]. The output voltage and output current of the TEM were measured by varying the external load resistance connected to the TEM.

Figure 7. TEHS made for lab test based on the numerical analysis.

The tested TEMs were of two types: 128A1030 (Peltron) and TK-1-3-S (HTRD). Generally, an I-V curve evaluating the power generation performance of the TEM is plotted on each temperature gradient to analyze performance with maximum output (Figure 8 and Table 2). The 128A1030 model and TK-1-3-S have maximum output when external resistance is approximately 1.4 Ω and 2.0 Ω, respectively. Both models have maximum output with external load resistance similar to the internal resistance value. While the railroad vehicle runs, the average temperature difference between the two sides of the TEM is approximately 15 °C, but the TK-1-3-S module has approximately 13% higher output performance. Therefore, the TK-1-3-S module is selected for the field test on a rolling stock with detailed design.

Table 2. Results of output performance test on TEMs with varying external load resistance, (a) results from 128A1030 (Peltron), (b) results from TK-1-3-S (HTRD).

(a)

ΔT (°C)	I at P_{max} (mA)	V at P_{max} (V)	R at P_{max} (Ω)	P_{max} (mW)
10	47.85	0.061	1.278	2.946
15	68.09	0.096	1.406	6.503
20	89.98	0.126	1.403	11.382
25	112.21	0.157	1.401	17.658
30	127.30	0.194	1.524	24.673

(b)

ΔT (°C)	I at P_{max} (mA)	V at P_{max} (V)	R at P_{max} (Ω)	P_{max} (mW)
10	41.40	0.078	1.892	3.296
15	62.02	0.118	1.900	7.295
20	75.14	0.165	2.197	12.701
25	100.42	0.201	2.001	19.879
30	111.83	0.246	2.197	27.969

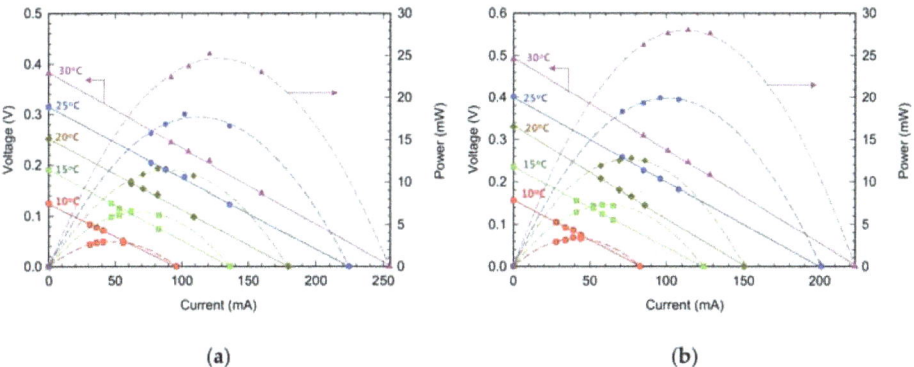

Figure 8. I–V characteristics of (a) 128A1030 (Peltron) and (b) TK-1-3-S (HTRD) TEMs for different temperature gradient.

4. Results and Discussion

Designed and manufactured for on-board test on a rolling stock based on the performance test of the TEM, a TEHS was installed on the surface of the axle bearing housing of the Korean next-generation high-speed railroad vehicle HEMU-430X (maximum speed 421.3 km/h), a prototype vehicle, between the axle bearing, heat source, and outdoor air. As shown in Figure 9, cooling fins with a TEM were manufactured to minimize the thermal leak of heat transfer from the heat source to the TEM and to maximize the temperature gradient by lowering the temperature of the low-temperature side of the TEM. The final shape of the cooling fins was designed and optimized considering the operation condition of the rolling stock and easy assembly of the protection frame. Figure 10 shows the TEHS with these design conditions applied to the test vehicle.

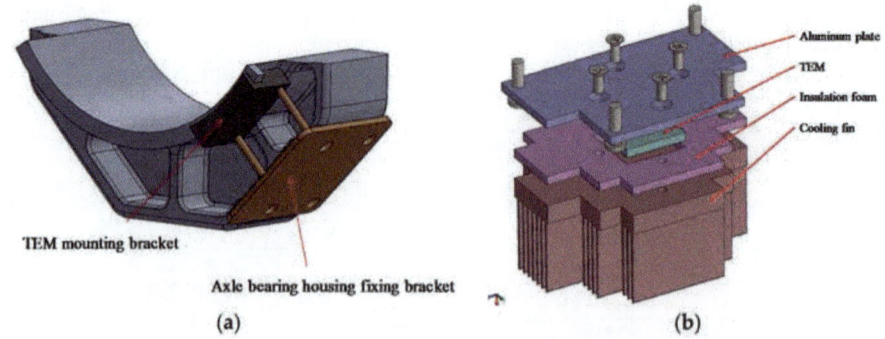

Figure 9. Design of (**a**) TEM mounting and fixing bracket and (**b**) cooling fin for applying on rolling

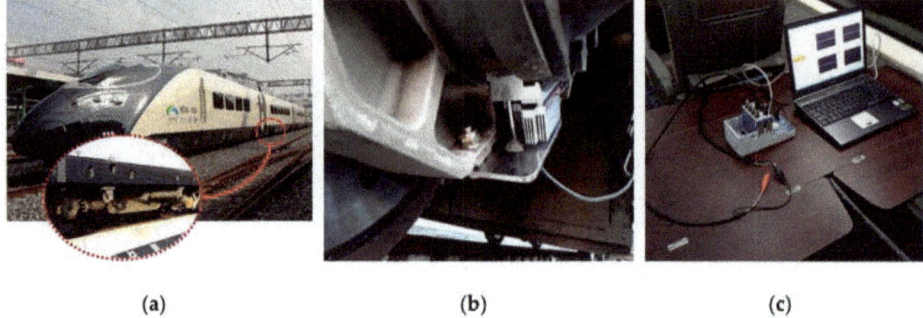

Figure 10. (**a**) HEMU-430X, Korean next generation high speed train, (**b**) TEHS mounted on axle bearing housing of HEMU-430X train, (**c**) On board data measurement setup.

The power generation test under rolling stock operation was performed. The test track was between Osong and Dongdaegu in Korea, and the maximum driving speed was 300 km/h. The output performance of the TEHS was measured in real time as shown in Figure 10 by a DAQ device (NI 9225, NI, Austin, TX, USA). At the same time, an individual thermocouple was installed on the axle bearing housing to measure the housing surface temperature in real time.

An open circuit voltage measured at the section between Osong and Dongdaegu is shown in Figure 11 together with the results of housing surface temperature and outdoor temperature. According to the conditions of railroad vehicle operation, the output voltage between the start of driving and approximately 1500 s (approximately 26 min) is unstable. That is because the initial driving speed is low, and the temperature of the axle bearing housing slowly increases, so stabilized power generation by TEM takes time. Actually, the surface temperature of the axle bearing housing increases after 800 s (approximately 13 min), and the maximum surface temperature reaches 38 °C at 2900 s without a temperature decrease. The open circuit voltage rapidly increases after 1600 s, and its maximum voltage is 0.4 V. As with the change in temperature, the output voltage also decreases after 2900 s when the railroad vehicle starts to decelerate to stop in Osong.

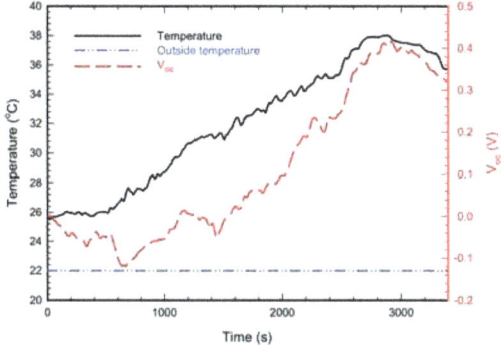

Figure 11. Measurement result of housing surface temperature and output voltage of the TEHS during Osong–Dongdaegu operation.

Based on the result of the field test, the estimated performance of the TEHS as a power source is presented in Figure 12. To harvest maximum power from the waste heat, it is important to select the optimal load resistance [37]. The result of the performance test on the TEM in Section 3 suggests that the optimal load resistance varies according to temperature conditions. Note, however, that configuring self-adjusting load resistance in accordance with temperature requires an additional power management circuit, and it may consume unnecessary electric power. Thus, the results in Figure 12 are obtained assuming constant optimal load resistance connected to the TEHS. The average temperature of the housing surface measured by the field test is approximately 36.8 °C, and the average temperature difference on both sides of the TEM is 14.8 °C. Therefore, accepting the result presented in Table 2b, the constant load resistance is assumed to be 1.9 Ω. Along with this condition, the output voltage, electric power, and charging energy with the assumed 100% charging efficiency of the TEHS are as shown in Figure 12. The TEHS has a maximum output voltage of approximately 0.192 V, its maximum power is approximately 19.3 mW, and its charged electric energy after railroad vehicle operation for 3400 s is approximately 16.6 J. Commercial low power sensors usually use 1.8 V or 3.3 V constant voltage source as a power supply. Therefore, the power management circuit is required to regulate power from the TEM, store electrical energy, and finally make constant voltage output. Assuming an electric energy conversion efficiency of 70%, which explains the energy loss of power management, approximately 11.6 J of electric energy can be stored for 3400 s. This amount of energy may continuously operate 30 commercial low-power temperature sensors (1.8 V–30 µA, PCT2202UK, NXP semiconductors, Eindhoven, The Netherland) or continuously operate 2 low-power 3-axis acceleration sensors (3.3 V–0.4 mA, MMA8491Q, NXP semiconductors, Eindhoven, The Netherland) for 3400 s.

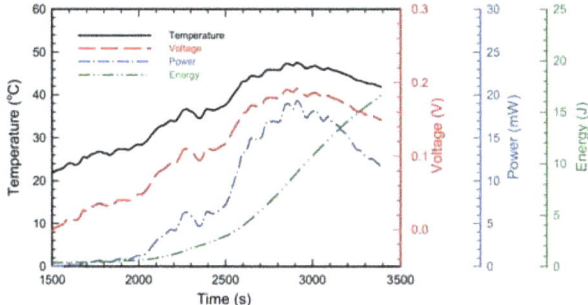

Figure 12. Expected power and energy output of the TEHS during Osong–Dongdaegu operation.

5. Conclusions

A thermoelectric energy harvesting system storing electrical energy by harvesting waste heat generated by railroad vehicles was designed and manufactured. To maximize the performance during the design process, a performance test on TEMs was performed, and cooling fins were optimized. A TEHS reflecting these optimized results and the compatibility with rolling stocks was manufactured and installed on the axle bearing housing of HEMU-430X, the Korean next-generation high-speed railroad vehicle. The thermoelectric power generation test using the temperature difference between the axle bearing housing and outdoor air during the rolling stock running yields up to 0.4 V of open circuit voltage. Based on the open circuit voltage and temperature data of the field test result, the expected generated power and stored energy can reach a maximum of 19.3 mW and 16.6 J, respectively. These are sufficient for operating a number of commercialized low-power sensors. Note, however, that the generated power is not sufficient at the very beginning of the railroad vehicle operation because it takes some time for the axle bearing housing to increase its temperature. Therefore, in order to use sensor nodes for monitoring the conditions of the running rolling stock in real time, further researches on highly functional, high-efficiency, low-power management are required. By applying the TEM to the rolling stock, the test of converting waste heat into useful electric energy was performed, and the waste heat of railroad vehicles was proven to be usable as a power source. In the future, a number of sensor nodes integrated with TEHS will be installed on various types of railroad vehicles, and the performance will be tested under various operation conditions.

Author Contributions: Conceptualization, K.C.; Methodology, K.C.; Validation, D.A.; Formal Analysis, D.A. and K.C.; Investigation, K.C.; Data Curation, K.C.; Writing-Original Draft Preparation, D.A. and K.C.; Writing-Review & Editing, D.A. and K.C.; Visualization, D.A. and K.C.; Supervision, K.C.; Project Administration, K.C.; Funding Acquisition, K.C.

Acknowledgments: This research was supported by a grant from R&D Program of the Korea Railroad Research Institute, Korea.

Conflicts of Interest: The authors declare no conflict of interest.

References

1. Garcia Marquez, F.P.; Schmid, F.; Collado, J.C. A reliability centered approach to remote condition monitoring. A railway points case study. *Reliab. Eng. Syst. Saf.* **2003**, *80*, 33–40. [CrossRef]
2. Kim, J. Infrared thermographic monitoring for failure characterization in railway axle materials. *J. Korean Soc. Nondestr. Test.* **2010**, *30*, 116–120.
3. Han, S.; Cho, S. Review of non-destructive evaluation technologies for rail inspection. *J. Korean Soc. Nondestr. Test.* **2011**, *31*, 398–413.
4. Choi, K.; Kim, J. A study for applying thermoelectric module in a bogie axle bearing. *Trans. Korean Soc. Mech. Eng. B* **2016**, *40*, 255–262. [CrossRef]
5. Koo, J.S.; Oh, H.S. A new derailment coefficient considering dynamic and geometrical effects of a single wheelset. *J. Mech. Sci. Technol.* **2014**, *28*, 3483–3498. [CrossRef]
6. Choi, S.; Kim, Y.H.; Lee, T.S. Case study on the incident caused by axle bearings of the railway vehicles. In Proceedings of the Korean Society for Railway Fall Conference, Jeju Island, Korea, 30–31 October 2014; Volume 10, pp. 1035–1040.
7. Choi, S.; Kim, M. A study on efficient rolling stock HBD monitoring method using EWMA technique. *J. Korea Acad. Industr. Coop. Soc.* **2017**, *18*, 609–617. [CrossRef]
8. Bhatia, D.; Lee, J.; Hwang, H.J.; Baik, J.M.; Kim, S.; Choi, D. Design of mechanical frequency regulator for predictable uniform power from triboelectric nanogenerators. *Adv. Energy Mater.* **2018**, *8*, 1702667. [CrossRef]
9. Zhao, X.; Wei, G.; Li, X.; Qin, Y.; Xu, D.; Tang, W.; Yin, H.; Wei, X.; Jia, L. Self-powered triboelectric nano vibration accelerometer based wireless sensor system for railway state health monitoring. *Nano Energy* **2017**, *34*, 549–555. [CrossRef]
10. Choi, W.; Choi, K.; Yang, G.; Kim, J.C.; Yu, C. Improving piezoelectric performance of lead-free polymer composites with high aspect ratio $BaTiO_3$ nanowires. *Polym. Test.* **2016**, *53*, 143–148. [CrossRef]

11. Choi, K.; Choi, W.; Yu, C.; Park, Y.T. Enhanced piezoelectric behavior of PVDF nanocomposite by AC dielectrophoresis alignment of ZnO nanowires. *J. Nanomater.* **2017**, *2017*, 6590121. [CrossRef]
12. Choi, K.; Ahn, D.; Boo, J.H. Influence of temperature gradient induced by concentrated solar thermal energy on the power generation performance of a thermoelectric module. *J. Korea Acad. Ind. Coop. Soc.* **2017**, *18*, 777–784. [CrossRef]
13. Tianchen, Y.; Jian, Y.; Ruigang, S.; Xiaowe, L. Vibration energy harvesting system for railroad safety based on running vehicles. *Smart Mater. Struct.* **2014**, *23*, 125046. [CrossRef]
14. Cahill, P.; Nuallain, N.A.N.; Jackson, N.; Mathewson, A.; Karoumi, R.; Pakrashi, V. Energy harvesting from train-induced response in bridges. *J. Bridge Eng.* **2014**, *19*, 04014034. [CrossRef]
15. Chu, S.; Majumdar, A. Opportunities and challenges for a sustainable energy future. *Nature* **2012**, *488*, 294–303. [CrossRef] [PubMed]
16. Chen, J.; Huang, Y.; Zhang, N.; Zou, H.; Liu, R.; Tao, C.; Fan, X.; Wang, Z.L. Micro-cable structured textile for simultaneously harvesting solar and mechanical energy. *Nat. Energy* **2016**, *1*, 16138. [CrossRef]
17. Chen, J.; Wang, J.L. Reviving vibration energy harvesting and self-powered sensing by a triboelectric nanogenerator. *Joule* **2017**, *1*, 480–521. [CrossRef]
18. Turner, J.A. A realizable renewable energy future. *Science* **1999**, *285*, 687–689. [CrossRef] [PubMed]
19. Bell, L.E. Cooling, heatng, generating power, and recovering waste heat with thermoelectric systems. *Science* **2008**, *321*, 1457–1461. [CrossRef] [PubMed]
20. Jacobson, M.Z. Review of solutions to global warming, air pollution, and energy security. *Energy Environ. Sci.* **2009**, *2*, 148–173. [CrossRef]
21. Wang, Z.L.; Chen, J.; Lin, L. Progress in triboelectric nanogenerators as a new energy technology and self-powered sensors. *Energy Environ. Sci.* **2015**, *8*, 2250–2282. [CrossRef]
22. Wang, E.H.; Zhang, H.G.; Fan, B.Y.; Ouyang, M.G.; Zhao, Y.; Mu, Q.H. Study of working fluid selection of organic Rankine cycle (ORC) for engine waste heat recovery. *Energy* **2011**, *36*, 3406–3418. [CrossRef]
23. Lin, Z.; Chen, J.; Yang, J. Recent progress in triboelectric nanogenerators as a renewable and sustainable power source. *J. Nanomater.* **2016**, *2016*, 5651613. [CrossRef]
24. Huang, K.D.; Tzeng, S. Development of a hybrid pneumatic-power vehicle. *Appl. Energy* **2005**, *8*, 47–59. [CrossRef]
25. Nas, B.; Berktay, A. Energy potential of biodiesel generated from waste cooking oil: An environmental approach. *Energy Sources Part B* **2007**, *2*, 63–71. [CrossRef]
26. Zhu, G.; Chen, J.; Zhang, T.; Jing, Q.; Wang, Z.L. Radial-arrayed rotary electrification for high performance triboelectric generator. *Nat. Commun.* **2014**, *5*, 3426. [CrossRef] [PubMed]
27. Chen, J.; Zhu, G.; Yang, W.; Jing, Q.; Bai, P.; Yang, Y.; Hou, T.C.; Wang, Z.L. Harmonic-resonator-based triboelectric nanogenerator as a sustainable power source and a self-powered active vibration sensor. *Adv. Mater.* **2013**, *25*, 6094–6099. [CrossRef] [PubMed]
28. Zhang, B.; Chen, J.; Jin, L.; Deng, W.; Zhang, L.; Zhang, H.; Zhu, M.; Yang, W.; Wang, J.L. Rotating-disk-based hybridized electromagnetic–triboelectric nanogenerator for sustainably powering wireless traffic volume sensors. *ACS Nano* **2016**, *10*, 6241–6247. [CrossRef] [PubMed]
29. Chen, J.; Yang, J.; Guo, H.; Li, Z.; Zheng, L.; Su, Y.; Wen, Z.; Fan, X.; Wang, Z.L. Automatic Mode Transition Enabled Robust Triboelectric Nanogenerators. *ACS Nano* **2015**, *9*, 12334–12343. [CrossRef] [PubMed]
30. Yang, J.; Chen, J.; Yang, Y.; Zhang, H.; Yang, W.; Bai, P.; Su, Y.; Wang, Z.L. Broadband vibrational energy harvesting based on a triboelectric nanogenerator. *Adv. Energy Mater.* **2014**, *4*, 1301322. [CrossRef]
31. Zhang, N.; Tao, C.; Fan, X.; Chen, J. Progress in triboelectric nanogenerator as self-powered smart sensors. *J. Mater. Res.* **2017**, *32*, 1628–1646. [CrossRef]
32. Armaroli, N.; Balzani, V. The future of energy supply: Challenges and opportunities. *Angew. Chem.* **2006**, *46*, 52–66. [CrossRef] [PubMed]
33. Chen, G.; Dresselhaus, M.S.; Dresselhaus, G.; Fleurial, J.P.; Caillat, T. Recent developments in thermoelectric materials. *Int. Mater. Rev.* **2003**, *48*, 45–66. [CrossRef]
34. Tritt, T.M.; Boettner, H.; Chen, L. Thermoelectris: Direct solar thermal energy conversion. *MRS Bull.* **2008**, *33*, 366–368. [CrossRef]
35. Majumdar, A. Thermoelectricity in Semiconductor nanostructures. *Science* **2004**, *303*, 777–778. [CrossRef] [PubMed]

36. Incropera, F.P. Convection Heat Transfer in Electronic Equipment Cooling. *J. Heat Transfer* **1988**, *110*, 1097–1111. [CrossRef]
37. Tan, Y.K.; Panda, S.K. Optimized wind energy harvesting system using resistance emulator and active rectifier for wireless sensor nodes. *IEEE T. Power Electr.* **2011**, *26*, 38–50. [CrossRef]

© 2018 by the authors. Licensee MDPI, Basel, Switzerland. This article is an open access article distributed under the terms and conditions of the Creative Commons Attribution (CC BY) license (http://creativecommons.org/licenses/by/4.0/).

Article

Effects of Embedded TiO$_{2-x}$ Nanoparticles on Triboelectric Nanogenerator Performance

Hyun-Woo Park [1,2,†], Nghia Dinh Huynh [1,†], Wook Kim [1], Hee Jae Hwang [1], Hyunmin Hong [2], KyuHyeon Choi [2], Aeran Song [2], Kwun-Bum Chung [2,*] and Dukhyun Choi [1,*]

[1] Department of Mechanical Engineering, Kyung Hee University, 1732 Deogyeong-daero, Giheung-gu, Yongin-Si, Gyeonggi-do 446-701, Korea; mnphwcj@gmail.com (H.-W.P.); hdnghia567@gmail.com (N.D.H.); choice124@hanmail.net (W.K.); chgw8584@hanmail.net (H.J.H.)
[2] Department of Physics and Semiconductor Science, Dongguk University, Seoul 100-715, Korea; ggm0218@naver.com (H.H.); cransia@hanmail.net (K.C.); aeransong@dongguk.edu (A.S.)
* Correspondence: kbchung@dongguk.edu (K.-B.C.); dchoi@khu.ac.kr (D.C.); Tel.: +82-2-2260-3187 (K.-B.C.); +82-31-201-3320 (D.C.)
† These authors are equally contributed.

Received: 23 July 2018; Accepted: 14 August 2018; Published: 17 August 2018

Abstract: Triboelectric nanogenerators (TENGs) are used as self-power sources for various types of devices by converting external waves, wind, or other mechanical energies into electric power. However, obtaining a high-output performance is still of major concern for many applications. In this study, to enhance the output performance of polydimethylsiloxane (PDMS)-based TENGs, highly dielectric TiO$_{2-x}$ nanoparticles (NPs) were embedded as a function of weight ratio. TiO$_{2-x}$ NPs embedded in PDMS at 5% showed the highest output voltage and current. The improved output performance at 5% is strongly related to the change of oxygen vacancies on the PDMS surface, as well as the increased dielectric constant. Specifically, oxygen vacancies in the oxide nanoparticles are electrically positive charges, which is an important factor that can contribute to the exchange and trapping of electrons when driving a TENG. However, in TiO$_{2-x}$ NPs containing over 5%, the output performance was significantly degraded because of the increased leakage characteristics of the PDMS layer due to TiO$_{2-x}$ NPs aggregation, which formed an electron path.

Keywords: TiO$_{2-x}$ nanoparticle; high dielectric constant; triboelectric nanogenerators; oxygen vacancy

---

1. Introduction

Triboelectric nanogenerators (TENGs) have demonstrated the ability to convert surrounding ambient mechanical energy into electricity based on the coupling effects of contact triboelectrification and electrostatic induction. However, the need for a low-cost method and an improvement in output performance remain a challenge in TENG fabrication. A practical application to harvest mechanical energy is also required. In general, the two key factors that significantly affect the output performance of TENGs are the tribo-material and the effective contact area of the friction layers. Another study showed that the surface charge density enhanced the triboelectric material and that the surface morphologies in micro/nanopatterning control the TENGs performance [1,2]. As shown in 2012 by Wang's group [3], TENGs operate in four basic modes: vertical contact-separation mode; lateral sliding mode; single electrode mode, and freestanding triboelectric-layer mode. Among these working modes, the vertical contact–separation mode was introduced first and has become the primary mode due to its simple operation, durability, stable and high-output performance. Basically, using the electrostatic charges created on the surfaces of two dissimilar materials when they are brought into contact, the tribo-charges can generate a potential drop when two surfaces are separated by an external force and, thus, drive electron flow between two electrodes connected on both sides of the tribo-material.

Recently, TENG development has become very common in many fields, such as energy harvesting, self-powered sensors, self-charging systems, or self-powered wearable electronic devices. TENGs has been investigated to obtain higher electricity output using surface charge density enhancement via ionized injection, which alters the effective contact area between the friction tribo-layers. This improvement is through its surface patterning or through new materials developed using functional group attachment. However, the simple, low-cost, and flexible structure of TENGs through embedded nanoparticles (NPs) is still being considered as a way to increase the capacitance of TENGs. By modifying the inside polymers through filling with nanoparticles and forming pores, many studies have reported the advantages of an embedded high dielectric constant material in polydimethylsiloxane (PDMS), such as $BaTiO_3$, $SrTiO_3$ [4,5], Au, or Ag nanoparticles [6], for improving TENG capacitance. In addition, the output performance of TENG can be increased by adjusting the distribution depth of the tribo-charges in the friction layer as described by Cui et al. [7]. C. Wu et al. [8] also demonstrated that the charge density in the friction layer is enhanced by utilizing a reduced graphene oxide (rGO) acting as electron-trapping sites [8]. These methods can lead to the higher output performance of TENGs, but are still limited in real applications due to the complexity, high-cost, and lack of both physical characteristic and analysis properties for understanding embedded NP behaviors inside the tribo-polymers layer. Moreover, oxygen vacancies in the oxide nanoparticles are electrically positive charges [9], which is an important factor that can contribute to the exchange and trapping of electrons when driving a TENG. In previous studies, however, the effect of oxygen vacancies within the tribo-material surface according to the oxide nanoparticle embedded ratio was not studied.

In this work, we report a significant improvement of TENG performance by considering TiO_{2-x} NPs embedded in PDMS as a function of weight ratio. Through this work, the optimal TiO_{2-x} NPs embedded PDMS is determined, and we investigated the correlation between TENG output behavior and the physical properties of the TiO_{2-x} NPs-embedded PDMS, such as the surface potential, dielectric properties, oxygen vacancy, and electronic structures. Different TiO_{2-x} NPs embedded according to PDMS weight ratios from 0 to 30% were chosen for tribo-layer fabrication. The results show a significantly enhanced TENG output upon the use of TiO_{2-x} NPs embedded PDMS as compared to a pristine PDMS tribo-layer. The optimal TiO_{2-x} NPs embedded PDMS with a 5% weight ratio exhibited the highest voltage of 180 V and current of 8.15 µA under a 5 N pushing force and 5 Hz pushing frequency. Furthermore, for practical applications, a portable windmill system integrated with our optimal TENGs using a low-cost, simple fabrication process involving plastic bottle waste is proposed.

2. Materials and Methods

2.1. The Fabrication Process of the PDMS Layer and TiO_{2-x} NPs Embedded PDMS Layer

Two hundred micrometers thick PDMS and TiO_{2-x} NPs embedded PDMS layers were fabricated by simple imprint lithography (SIL) using a doctor–blade and a Teflon template as a mold. The Teflon template was cleaned with acetone, ethanol, and deionized (DI) water, consecutively, and then gently dried with dry nitrogen (N_2) gas. To prevent the PDMS layer from sticking to the mold, we treated the surface of the Teflon mold with heptadecafluoro-1, 1, 2, 2-tetrahydrodecyltrichlorosilane (HDFS) to make its surface hydrophobic. The PDMS was prepared as a mixture of base resin and curing agent (Sylgard 184 A: Sylgard 184 B, Dow Corning Co., Midland, MI, USA) with a weight ratio of 10:1. After a degassing process under vacuum for approximately 30 min, the PDMS mixture was then laminated to the Teflon mold by a doctor–blade technique and cured at 80 °C for six hours. The 200 µm thick PDMS layer was then carefully peeled off the Teflon mold. The TiO_{2-x} NPs-embedded PDMS layer was fabricated similarly, except the embedding process of TiO_{2-x} NPs (US Research Nanomaterials, Inc. TiO_{2-x} Nanoparticles, rutile, high purity, 99.9+%) utilized different mass ratios from 5 to 30%, dispersed into the PDMS mixture.

2.2. Fabrication and Output Measurement of the TENGs

To design the vertical contact–separation mode TENG, the top electrode was prepared with a 3×3 cm^2, 80 µm-thick commercial aluminum foil, which was cleaned by ethanol and dried using N$_2$ gas. The aluminum foil was then attached to the polylactic acid (PLA) substrate with double-sided foam tape. Subsequently, to prepare the TENG bottom electrode, the PDMS layer and TiO$_{2-x}$ NPs embedded PDMS layer were carefully laminated onto the aluminum plate. Figure S1 (Supporting Information) shows a schematic of the fabrication process for TiO$_{2-x}$ NPs embedded TENG.

2.3. Analysis of the TiO$_{2-x}$ NPs-Embedded TENG

The physical structure of TiO$_{2-x}$ nanoparticles was analyzed using X-ray diffraction (XRD). The surface morphology characteristics of the PDMS layer and TiO$_{2-x}$ NPs-embedded PDMS layer were examined using scanning electron microscopy (SEM) and atomic force microscopy (AFM). To measure the TENG output performance, a force was applied using a pushing tester. The TENG output voltage was measured using a Tektronix MDO3052 (Tektronix, Inc., Beaverton, OR, USA) mixed-domain oscilloscope with an input impedance of 40 MΩ, and output current measurements were captured using a Stanford Research Systems SR570 low-noise current preamplifier (Stanford Research System, Sunnyvale, CA, USA) connected to the Tektronix MDO3052 mixed domain oscilloscope. The TENG performance was measured at a working frequency of 5 Hz and a force of 5 N; the gap between the two electrodes was 10 mm. The chemical bonding states and composition of PDMS and TiO$_{2-x}$ NPs-embedded PDMS layers were also examined by X-ray photoelectron spectroscopy (XPS) with a pass energy of 20 eV, using a monochromatic Al Kα source. The change in the unoccupied state in the conduction band according to the TiO$_{2-x}$ percentage was investigated using X-ray absorption spectroscopy (XAS). XAS experiments were performed at the 10 D beamline in the Pohang Accelerator Laboratory (PAL), Korea. In addition, the triboelectric charges of PDMS, according to the TiO$_{2-x}$ embedded weight ratio, were quantitatively studied using Kelvin probe force microscopy (KPFM, N8-NEOS, Bruker, Billerica, MA, USA) and contact potential difference (CPD) measurements.

3. Results and Discussion

The proposed TENG was operated using a simple model of contact–separation, as depicted in Figure 1a. The TiO$_{2-x}$ NPs-embedded PDMS layer was laminated on the Al electrode as a negative tribo-material. When the TiO$_{2-x}$ NPs-embedded PDMS layer comes into contact with the top layer Al electrode (i.e., positive tribo-material) as well, the TiO$_{2-x}$ NPs-embedded PDMS layer becomes negatively charged while the Al electrode is positively charged because of the different electron affinities. During the release process, potential is created between the two opposite electrostatic charges. The potential across these electrodes leads to an electron flow through the external circuit from the bottom Al electrode to the top to obtain electrical equilibrium. After full separation, the electrodes revert back to their original states, and at last, the tribo-charge distribution reaches electrical equilibrium. Applying a continuous external pushing force to the TENGs, repeated TENG operation can be observed. Since the titanium oxide materials have various physical structures, such as anatase, rutile, and bookite, and their electrical, optical, and physical properties depend on their physical structure [10], XRD patterns of the TiO$_{2-x}$ NPs were investigated, as shown in Figure 1b. The TiO$_{2-x}$ NPs represent the typical polycrystalline rutile structure with diffraction peaks of (110), (101), (200), (101), (210), (211), (220), (002), (310), (301), and (112), and the calculated particle size was about 20 nm through main XRD peak of (110). [11,12] The energy band diagram of the PDMS layer and TiO$_{2-x}$ NPs-embedded PDMS layer with 5% and 30% are shown in Figure 1c. As TiO$_{2-x}$ NP increased, the work function difference between the two contact materials is reduced from 2.61 eV to 0.72 eV. This implies that the transport of electrons from the top Al electrode can be reduced.

Figure 1d shows the scanning electron microscope (SEM, compact SEM GENESIS-1000, Emcrafts, Korea) images of the PDMS layer and TiO$_{2-x}$ NPs embedded PDMS layer with 5% and 30%. The surface morphology of these PDMS and TiO$_{2-x}$ NPs-embedded PDMS layers are uniformly wrinkle-like and similar, except for the higher weight ratio TiO$_{2-x}$ NPs-embedded PDMS layers. The average roughness of these TiO$_{2-x}$ NPs-embedded PDMS layers is confirmed by the AFM image, and the mean surface roughness of the PDMS layer and TiO$_{2-x}$ NPs-embedded PDMS layer with 5% and 30% was 17.8, 22.1, and 19.8 nm, respectively. (Not shown here).

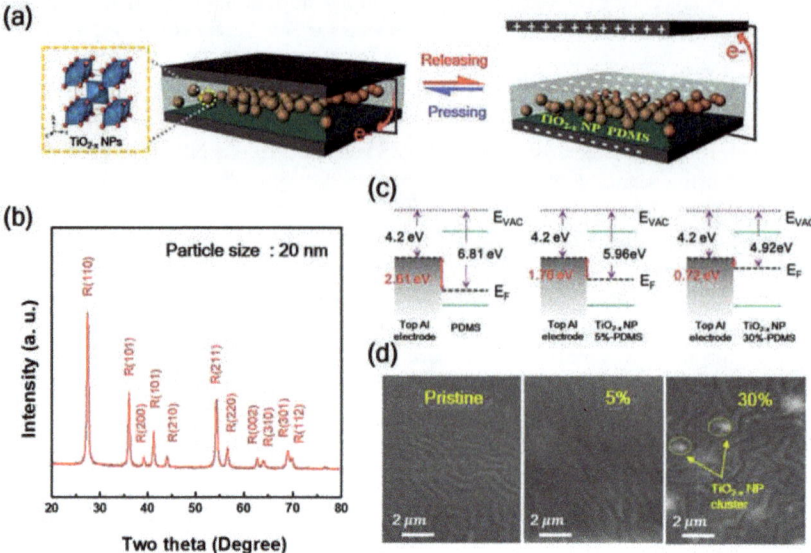

Figure 1. (a) Schematic of the vertical contact–separation mode triboelectric nanogenerators (TENG) with TiO$_{2-x}$ NPs embedded in polydimethylsiloxane (PDMS) as friction layers; alternating current is generated upon pressing and releasing; (b) X-ray diffraction (XRD) spectra of the TiO$_{2-x}$ nanoparticles (NPs); (c) The energy band diagram, and (d) Scanning electron microscope (SEM) image of PDMS as a function of the TiO$_{2-x}$ NPs embedded weight ratio.

Figure 2a,b show the output voltage and current characteristics of the TENG without/with TiO$_{2-x}$ NPs as a function of weight ratio (measured under a relative humidity of ~55% and room temperature of 27 °C). As the weight ratio of TiO$_{2-x}$ NPs increased from 0% (pristine PDMS) to 5%, the output voltage and current increased significantly from 112 V and 3.5 µA to 180 V and 8.15 µA, respectively. However, at weight ratios greater than 5%, the output voltage and current drastically decreased to 38.7 V and 0.6 µA, respectively. Moreover, to examine the effect of the pushing force on the output performance of the TENG device with pristine PDMS, 5% (optimal weight ratio) and 30% TiO$_{2-x}$ NPs-embedded PDMS, the output voltage and current of the TENG were measured and analyzed at different pushing forces ranging from 5 to 20 N and pushing frequency of 5 Hz (Figure S2—Supporting Information). It clearly indicates that the peak voltage and current of the TENG were increased from 48.2 V and 1.6 µA to 91.8 V and 2.9 µA for the pristine TENG, from 64.9 V and 3.8 µA to 126.8 V and 6.1 µA for 5%, and that of 30% TiO$_{2-x}$ NPs-embedded PDMS based TENG increased from 30.6 V and 0.9 µA to 58.7 V and 2.1 µA by enhancing the applied pushing forces from 5 N to 20 N, respectively. The highest output voltage and current values of 126 V and 6.1 µA were obtained under the externally applied force of 20 N. This result is mainly attributed to the surface potential dramatically changing among these kind of TENGs. This was especially true for the 30% TiO$_{2-x}$ NPs-embedded PDMS which became a positively charged material, leading to the drop of the TENG output performance, even though a strong applied pushing force. In addition, the trend of the TENG output among pristine, 5%, and 30% is the same as the measurement at 5 N and 5 Hz shown in Figure 2. It was noted that the surrounding conditions were 75% relative humidity and 32 °C room temperature for the different applied forces measurements. Figure 2c shows the effect of voltage and current characteristics of the TENG as a function of external load resistance ranging from 0.1 to 1000 MΩ. These measurements were performed under a pushing force and frequency of 5 N and 5 Hz, respectively. As shown in Figure 2c, the TENG voltage increased by increasing the resistance from 0.1 to 200 MΩ, while the current value followed the opposite trend. Afterward, both the voltage and current became saturated at very high load resistances (i.e., 200–1000 MΩ), and the TENG exhibited the highest voltage and lowest current values ranging from 230.6–260 V and 4.8–2.6 µA. In addition, the average power density (P_{avg}) was calculated by the following the equation:

$$P_{avg} = \frac{VI}{S} \qquad (1)$$

where V and I are the output peak voltage and current values of the TENG at various load resistances, respectively, and S is the active area (i.e., 9 cm^2) of the TENG. Figure 2d shows the effect of the TENG output power density as a function of external load resistance. Following Figure 2d, the P_{avg} values of TENG increased by increasing the load resistance and then decreased at a relatively high load resistance. Specifically, at the moderate load resistance of 40 MΩ, the TENG exhibited a maximum power density of 1.84 W.m^{-2}, under the pushing force and frequency of 5 N and 5 Hz, respectively. More discussion is provided below regarding the output performance behavior of the TENGs and the PDMS properties according to the TiO$_{2-x}$ NP embedding ratio.

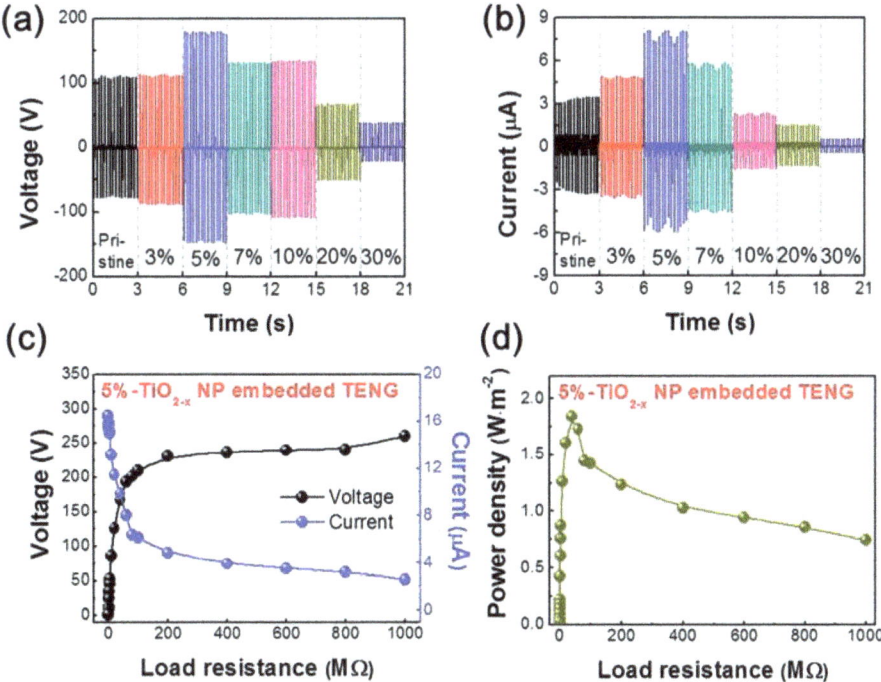

Figure 2. Effects of the output power of TENGs caused by embedded TiO$_{2-x}$ NPs. (**a**) Output voltage and (**b**) current of the TiO$_{2-x}$ NP embedded TENGs as a function of the TiO$_{2-x}$ NPs embedded weight ratio (measured under a relative humidity of ~55%). Influence of external load resistance on the (**c**) output voltage/current and (**d**) power density of the TENG under a pushing force of 5 N and a pushing frequency of 5 Hz.

Previous studies show that as the embedded weight ratio of NPs with high dielectric constants increases, the TENG output performance was improved because of the increasing dielectric constant of the tribo-material [13]. Our results show the same trends as previous studies, as shown in Figure 3a. In addition, the dielectric constant according to frequency does not change significantly. However, we provide below not only an increase in the dielectric constant of the tribo-material but also suggest an additional mechanism based on the role of oxygen vacancy in TiO_{2-x} NP embedded PDMS to effectively enhance TENG performance. To investigate changes of chemical bonding state within the PDMS layer according to the amount of embedded TiO_{2-x} NPs, the oxygen (O) 1s spectra of PDMS and TiO_{2-x} NPs 5% embedded PDMS layers were observed via XPS, as shown in Figure 3b. For the detailed analysis of chemical bonding states, the O 1s spectra were carefully normalized and de-convoluted with three different Gaussian peaks as indexed with O1, O2, and O3 centered at 530.9 eV, 531.8 eV, and 532.9 eV, respectively. Each of the representatively assigned peaks from a low binding energy is related to the oxygen state in the metal-oxide lattices, the oxygen-deficient state, and chemisorbed or dissociated oxygen states, or OH^- impurities, respectively [14,15]. Among them, the relative area of the oxygen-deficient peak (O2) is dramatically increased from 18.80% to 24.63% after TiO_{2-x} NPs embedding, which is remarkable. Moreover, we measured and de-convoluted O 1s' peak of other films that have different concentration, as shown in Figure S3. In the result, the oxygen-deficient states were similar regardless of increases of TiO_{2-x} concentration. This saturation can be observed due to the formation of TiO_{2-x} clusters in the composite film (see Figure 1d). The TiO_{2-x} nanoparticles in a cluster act like a single particle. This means that the number of active oxygen deficient states are as same as a single TiO_{2-x} nanoparticle. Due to the oxygen vacancies being electrically positively charged, the increment of oxygen vacancies of the negative tribo-material surface could induce electron flow from the top electrode by the attraction force, as shown in Figure 3c. Therefore, the TiO_{2-x} NPs-embedded PDMS layer has a larger oxygen-deficient state by oxygen vacancies, which can contribute to an enhanced TENG performance. To theoretically understand the output performance enhancement of the TiO_{2-x} NPs-embedded TENG, a COMSOL Multiphysics (5.0, COMSOL, Inc.) simulation was conducted. Figure 3d shows the corresponding results for the triboelectric potential distributions of pristine TENGs and TiO_{2-x} NPs-embedded TENG, respectively. The details of the COMSOL Multiphysics simulation parameters and equations are provided in our previous study, Reference [16]. As a result, the COMSOL simulations showed that the 5% TiO_{2-x} NPs-embedded TENG can exhibit higher performance than the pristine TENG.

To elucidate the cause of the significantly decreased output performance of TENG containing over 5% embedding TiO_{2-x} NPs, the correlation between the electronic structures of the TiO_{2-x} NPs-embedded PDMS and the energy conversion mechanism of the TENG were investigated. Figure 4a shows the hybridized molecular orbital structures in the conduction band of the TiO_{2-x} NPs-embedded PDMS as a function of the weight ratio of TiO_{2-x} NPs measured by XAS. The normalized intensities of the oxygen K-edge spectra of pristine PDMS and TiO_{2-x} NPs-embedded PDMS layers directly reflect the unoccupied hybridized states by the transition of electrons from the occupied O 1s state [17]. Figure 4a co-plots the normalized O-K edge XAS spectrum of TiO_{2-x} NPs (gray lines) for easy comparison with the changes in the TiO_{2-x} NPs-embedded PDMS layer spectra. In particular, XAS analysis is a surface sensitive analytical method that measures the current flowing on the sample surfaces after incident X-ray exposure. Therefore, this is suitable for analyzing tribo-materials whose output performance changes according to surface conditions. The XAS spectra of TiO_{2-x} NPs are located in the two strong resonance peaks at 531.5 and 534 eV [18]. With an increased weight ratio of TiO_{2-x} NPs from 0 to 30%, the electronic structure of PDMS is clearly changed by the addition of the electronic structure of TiO_{2-x} NPs. These changes in the tribo-material surface can cause changes in the tribo-series, resulting in reduced tribo-electrification effects. For the experimental demonstration of our claim, KPFM measurements were carried out to detect the surface potential of different surfaces according to the weight ratio of TiO_{2-x} NPs, as shown in Figure 4b. As the weight ratio of TiO_{2-x} NPs increased from 0 to 30%, the average surface potential dramatically changed. Specifically, the

TiO$_{2-x}$ NPs 30% sample became a positively charged material. Moreover, considering the SEM image in Figure 1d, the electron path can be formed due to NPs clustering in the case of the TENG containing over 5% embedding TiO$_{2-x}$ NPs. The electron path formed by the NPs cluster can move the electrons induced on the surface and increase the leakage current, as shown in Figure S4 (supporting information), which may be a reasonable cause to reduce TENG performance above 5% embedding TiO$_{2-x}$ NPs, as shown in Figure 4c. To demonstrate this with a practical application, we designed a windmill system to harvest the wind energy utilizing the pristine TENG and optimal TENG devices (TiO$_{2-x}$ NPs 5% embedded TENG), as shown in Figure 5a. In the presence of wind flow, the windmill blades are forced to rotate and lead to the rotation of a quad nose cam. Subsequently, the quad nose cam transmits its rotation motion to linear motion on the top plate of the TENG. Resulting from that motion, the TENG operates in the vertical direction via contact and separation mode repeatedly (Figure 5b), which produces the TENG electrical output. Moreover, the relation of wind speed and output performance of the windmill-integrated TENG was also analyzed at various wind speeds, as shown in Figure 5c. The measured voltage is clearly shown in the output performance of both the pristine and optimal TENGs, which show linear enhancement upon increasing the wind speed applied to the windmill blades. The peak voltage values of the windmill attached to pristine TENG at the wind speeds of ~10, 15, 20, 25, 30, and 35 are 70, 92, 114, 130, 182, and 190, and that for the windmill integrated with an optimal TENG are 94, 172, 220, 268, 304, and 304, respectively. The output data in Figure 5c also show a linear enhancement regarding the number of pushing cycles, i.e., pushing frequency and force on the top plate of the TENG as the wind speed increases. With faster wind speeds, the output performance of the windmill TENG can be enhanced. The results clearly demonstrate that our windmill TENG system can efficiently harvest wind energy.

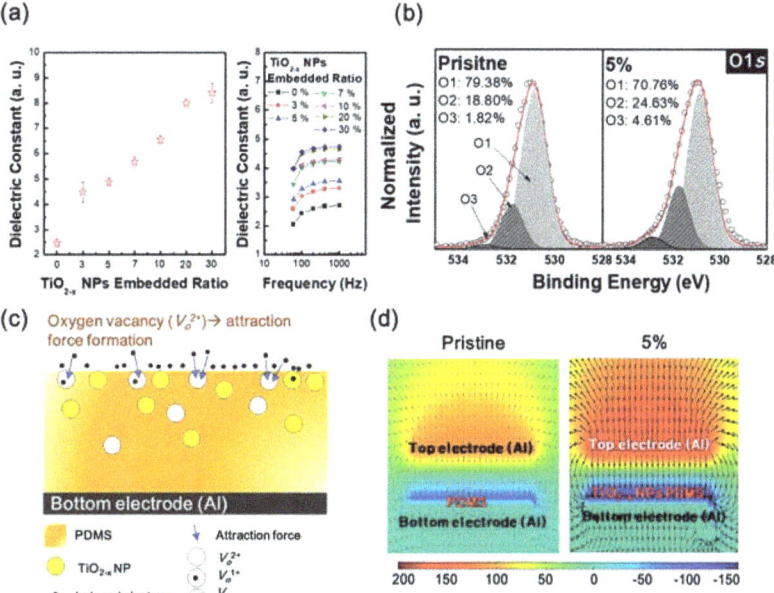

Figure 3. (a) Dielectric constant of the TiO$_{2-x}$ NPs embedded PDMS layers according to the frequency with different TiO$_{2-x}$ NPs weight ratios; (b) De-convoluted X-ray photoelectron spectroscopy (XPS) spectra of oxygen (O) 1s states of pristine PDMS layer and TiO$_{2-x}$ NPs 5% embedded PDMS layer; (c) Schematic of the electron attraction mechanism according to oxygen vacancies of the PDMS surface, and (d) COMSOL simulation results of the pristine TENG and TiO$_{2-x}$ NPs 5% embedded TENG.

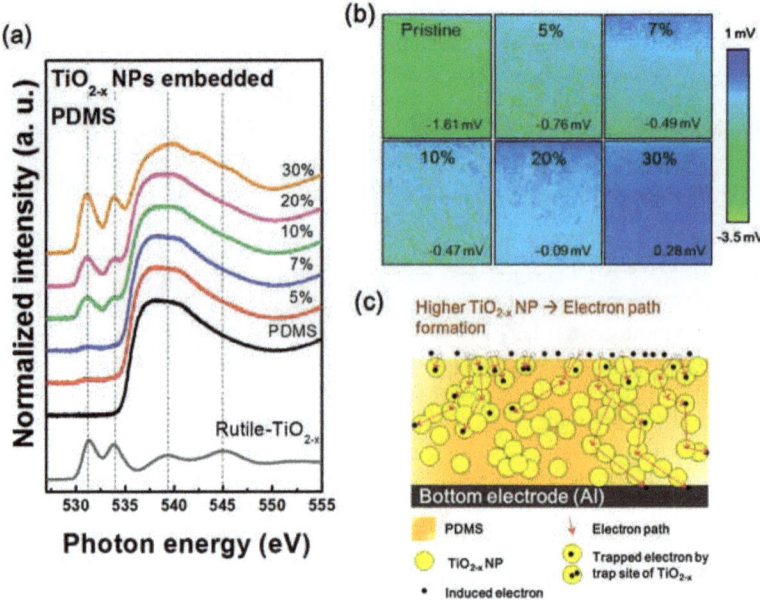

Figure 4. (**a**) O-K edge X-ray absorption spectroscopy (XAS) spectra and (**b**) Kelvin probe force microscopy (KPFM) images of PDMS layers as a function of the TiO$_{2-x}$ NPs embedded weight ratios; (**c**) Schematic of the electron leakage mechanism according to electron path formation.

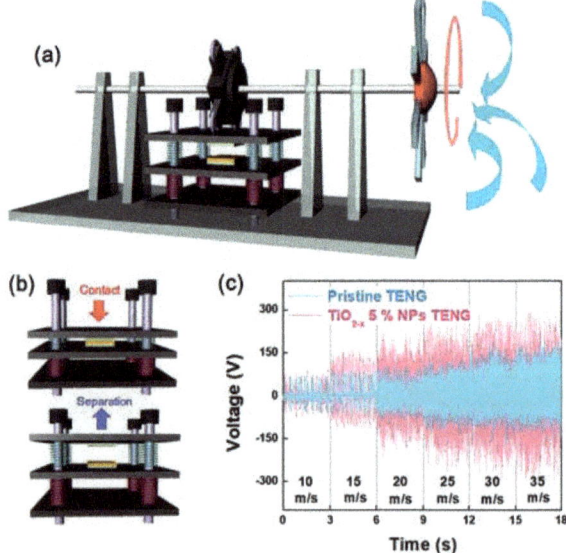

Figure 5. (**a**) Schematic of the windmill system and (**b**) TENG operation in the vertical direction repetitively switching between the contact and separation modes; (**c**) Measured voltage of the TiO$_{2-x}$ NPs 0% (pristine TENG) and 5% embedded TENG at different wind speeds applied to the windmill system.

4. Conclusions

In summary, we demonstrated a facile approach to improve the output performance of TENGs by embedding high dielectric TiO_{2-x} nanoparticles in the PDMS layer. The output performance of the TiO_{2-x} NPs embedded (5%) TENG showed the highest output voltage and current of 180 V and 8.15 µA, respectively, which is strongly related to the change of oxygen vacancies on the PDMS surface as well as the increased dielectric constant. Since the oxygen vacancies are positively charged, the increment of oxygen vacancies of the negative tribo-material surface could induce electrons from the top electrode by attraction force. Therefore, the TiO_{2-x} NPs-embedded PDMS layer has larger oxygen vacancies, which can contribute to the enhanced TENG performance. In contrast, with TiO_{2-x} NPs embedded samples over 5%, the output voltage and current were drastically decreased to 38.7 V and 0.6 µA, respectively. This significant degradation of output performance is due to variation of the tribo-series, resulting in reduced tribo-electrification effects. In addition, the electron path could be formed due to TiO_{2-x} NPs clustering, which can move the electrons induced on the surface thereby increasing leakage current. Last, our optimal TENG was also presented as a practical application for wind energy harvesting and could obtain 424 V of the peak output voltage via the proposed windmill.

Supplementary Materials: The following are available online at http://www.mdpi.com/2072-666X/9/8/407/s1, Figure S1: Schematic of the fabrication process for TiO_{2-x} NPs embedded TENG, Figures S2: TENG performance according to the different applied input forces, Figures S3: Change of oxygen deficient states with various TiO_{2-x} NP concentrations, Figures S4: Leakage current characteristics of PDMS layer as a function of TiO_{2-x} NPs embedded ratio.

Author Contributions: H.-W.P. and N.D.H. designed the experimental concept and wrote the main manuscript. W.K., H.J.H., H.H, K.C. and A.S. discussed the experimental results and commented on the theoretical mechanisms. All authors reviewed the manuscript. The project was guided by K.-B.C. and D.C.

Funding: This work was supported by a National Research Foundation of Korea (NRF) grant funded by the Korean government (MSIP) (No.2016R1A4A1012950), (No. 2017R1A6A3A11029892), (No. 2017R1D1A1B03032375), and (NRF-2015K2A2A7056357).

Conflicts of Interest: The authors declare no conflicts of interest.

References

1. Dudem, B.; Huynh, N.D.; Kim, W.; Kim, D.H.; Hwang, H.J.; Choi, D.; Yu, J.S. Nanopillar-array architectured PDMS-based triboelectric nanogenerator integrated with a windmill model for effective wind energy harvesting. *Nano Energy* **2017**, *42*, 269–281. [CrossRef]
2. Kim, D.; Jeon, S.B.; Kim, J.Y.; Seol, M.L.; Kim, S.O.; Choi, Y.K. High-performance nanopattern triboelectric generator by block copolymer lithography. *Nano Energy* **2015**, *12*, 331–338. [CrossRef]
3. Fan, F.R.; Tian, Z.Q.; Wang, Z.L. Flexible triboelectric generator. *Nano Energy* **2012**, *1*, 328–334. [CrossRef]
4. Chen, J.; Guo, H.; He, X.; Liu, G.; Xi, Y.; Shi, H.; Hu, C. Enhancing performance of triboelectric nanogenerator by filling high dielectric nanoparticles into sponge PDMS film. *ACS Appl. Mater. Interfaces* **2016**, *8*, 736–744. [CrossRef] [PubMed]
5. Seung, W.; Yoon, H.J.; Kim, T.Y.; Ryu, H.; Kim, J.; Lee, J.H.; Lee, J.H.; Kim, S.; Park, Y.K.; Park, Y.J.; et al. Boosting Power-Generating Performance of Triboelectric Nanogenerators via Artificial Control of Ferroelectric Polarization and Dielectric Properties. *Adv. Energy Mater.* **2017**, *7*, 1600988. [CrossRef]
6. Chun, J.; Kim, J.W.; Jung, W.S.; Kang, C.Y.; Kim, S.W.; Wang, Z.L.; Baik, J.M. Mesoporous pores impregnated with Au nanoparticles as effective dielectrics for enhancing triboelectric nanogenerator performance in harsh environments. *Energy Environ. Sci.* **2015**, *8*, 3006–3012. [CrossRef]
7. Cui, N.; Gu, L.; Lei, Y.; Liu, J.; Qin, Y.; Ma, X.; Hao, Y.; Wang, Z.L. Dynamic behavior of the triboelectric charges and structural optimization of the friction layer for a triboelectric nanogenerator. *ACS Nano* **2016**, *10*, 6131–6138. [CrossRef] [PubMed]
8. Wu, C.; Kim, T.W.; Choi, H.Y. Reduced graphene-oxide acting as electron-trapping sites in the friction layer for giant triboelectric enhancement. *Nano Energy* **2017**, *32*, 542–550. [CrossRef]

9. Park, H.W.; Song, A.; Choi, D.; Kim, H.-J.; Kwon, J.-Y.; Chung, K.-B. Enhancement of the Device Performance and the Stability with a Homojunction-structured Tungsten Indium Zinc Oxide Thin Film Transistor. *Sci. Rep.* **2017**, *7*, 11634. [CrossRef] [PubMed]
10. Reyes-Coronado, D.; Rodríguez-Gattorno, G.; Espinosa-Pesqueira, M.E.; Cab, C.; de Coss, R.; Oskam, G. Phase-pure TiO_2 nanoparticles: Anatase, brookite and rutile. *Nanotechnology* **2008**, *19*, 145605. [CrossRef] [PubMed]
11. Yan, J.; Wu, G.; Guan, N.; Li, L.; Lib, Z.; Cao, X. Understanding the effect of surface/bulk defects on the photocatalytic activity of TiO_2: Anatase versus rutile. *Phys. Chem. Chem. Phys.* **2013**, *15*, 10978–10988. [CrossRef] [PubMed]
12. Sakurai, K.; Mizusawa, M. X-ray Diffraction Imaging of Anatase and Rutile. *Anal. Chem.* **2010**, *82*, 3519–3522. [CrossRef] [PubMed]
13. Fang, Z.; Chan, K.H.; Lu, X.; Tana, C.F.; Ho, G.W. Surface texturing and dielectric property tuning toward boosting of triboelectric nanogenerator performance. *J. Mater. Chem. A* **2018**, *6*, 52–57. [CrossRef]
14. Park, H.-W.; Song, A.; Kwon, S.; Ahn, B.D.; Chung, K.-B. Improvement of device performance and instability of tungsten-doped InZnO thin-film transistor with respect to doping concentration. *Appl. Phys. Exp.* **2016**, *9*, 111101. [CrossRef]
15. Kim, B.K.; Park, J.-S.; Kim, D.H.; Chung, K.-B. Semiconducting properties of amorphous GaZnSnO thin film based on combinatorial electronic structures. *Appl. Phys. Lett.* **2014**, *104*, 182106. [CrossRef]
16. Park, H.-W.; Huynh, N.D.; Kim, W.; Lee, C.; Nam, Y.; Lee, S.; Chung, K.-B.; Choi, D. Electron blocking layer-based interfacial design for highly-enhanced triboelectric nanogenerators. *Nano Energy* **2018**, *50*, 9–15. [CrossRef]
17. Kamada, M.; Hideshima, T.; Azuma, J.; Yamamoto, I.; Imamura, M.; Takahashi, K. Occupied and unoccupied electronic structures of an L-cysteine film studied by core-absorption and resonant photoelectron spectroscopies. *AIP Adv.* **2016**, *6*, 045306. [CrossRef]
18. Park, J.; Ok, K.-C.; Ahn, B.D.; Lee, J.H.; Park, J.-W.; Chung, K.-B.; Park, J.-S. Molecular orbital ordering in titania and the associated semiconducting behaviour. *Appl. Phys. Lett.* **2011**, *99*, 142104. [CrossRef]

 © 2018 by the authors. Licensee MDPI, Basel, Switzerland. This article is an open access article distributed under the terms and conditions of the Creative Commons Attribution (CC BY) license (http://creativecommons.org/licenses/by/4.0/).

Communication

Mechanical Fatigue Resistance of Piezoelectric PVDF Polymers

Youn-Hwan Shin [1], Inki Jung [1,2], Hyunchul Park [3], Jung Joon Pyeon [1], Jeong Gon Son [4], Chong Min Koo [1,3], Sangtae Kim [2,*] and Chong-Yun Kang [1,2,*]

1. KU-KIST Graduate School of Converging Science and Technology, Korea University, Seoul 02841, Korea; shw9106@kist.re.kr (Y.-H.S.); nasa1011@kist.re.kr (I.J.); 114342@kist.re.kr (J.J.P.); koo@kist.re.kr (C.M.K.)
2. Center for Electronic Materials, Korea Institute of Science and Technology, Seoul 02792, Korea
3. Materials Architecturing Research Center, Korea Institute of Science and Technology, Seoul 02792, Korea; hpark@kist.re.kr
4. Photo-Electronic Hybrid Research Center, Korea Institute of Science and Technology, Seoul 02792, Korea; jgson@kist.re.kr
* Correspondence: stkim@kist.re.kr (S.K.); cykang@kist.re.kr (C.-Y.K.); Tel.: +82-02-958-6623 (S.K.); +82-02-958-6722 (C.-Y.K.)

Received: 18 September 2018; Accepted: 3 October 2018; Published: 4 October 2018

Abstract: The fatigue resistance of piezoelectric PVDF has been under question in recent years. While some report that a significant degradation occurs after 10^6 cycles of repeated voltage input, others report that the reported degradation originates from the degraded metal electrodes instead of the piezoelectric PVDF itself. Here, we report the piezoelectric response and remnant polarization of PVDF during 10^7 cycles of repeated compression and tension, with silver paste-based electrodes to eliminate any electrode effect. After applying repeated tension and compression of 1.8% for 10^7 times, we do not observe any notable decrease in the output voltage generated by PVDF layers. The results from tension experiments show stable remnant polarization of 5.5 $\mu C/cm^2$, however, the remnant polarization measured after repeated compression exhibits a 7% decrease as opposed to the tensed PVDF. These results suggest a possible anisotropic response to stress direction. The phase analyses by Raman spectroscopy reveals no significant change in the phase content, demonstrating the fatigue resistance of PVDF.

Keywords: ferroelectric; PVDF; piezoelectric; mechanical fatigue resistance; remnant polarization

1. Introduction

Piezoelectric polymers such as polyvinylidene fluoride (PVDF) have been studied extensively in terms of their piezoelectric properties, with applications ranging from roadway energy harvesting to film-based speakers and biomedical applications [1–3]. Some of these applications require excellent mechanical and polarization fatigue resistance. For instance, energy harvesters for roadway applications need to withstand mechanical shocks on the order of 10^8 times, assuming the daily traffic of 13,000 vehicles and the road's life expectancy of 10 years [4,5]. The PVDF layers for film speakers require similar resistance against polarization fatigue, assuming the speaker plays 500 Hz audio sound for 56 h [2]. Also, the PVDF-based generator exhibits stable electric output with no degradation after 5 days of in vivo operation (mouse) or 1.512×10^8 times mechanical deformation [3].

However, measuring the fatigue resistance of PVDF remains difficult, since the metal electrodes on flexible PVDF easily deform under repeated strain and affect the measurements. For instance, Zhu et al. reported in 2006 that the remnant and saturation polarization of PVDF copolymered with trifluoroethylene (PVDF-TrFE) decays by 40% when repeated electrical bias up to 10^6 times is applied [6]. In a recent paper, however, Zhao et al. showed that this decay in polarization originates from partial delamination of metal electrodes during fast bias switching instead of the degradation of PVDF layer itself [7]. In direct application of strain via bending, similar degradation behavior of metal electrodes is expected. For example, Sim et al. showed that polymer-supported metal thin films undergo crack propagation, debonding or delamination, leading to increased resistance under repeated mechanical strain [8,9]. Since the elastic limit of metal thin films falls far below that of flexible polymers, the accumulating plastic deformation on metal thin films inevitably results in the inaccurate assessment of PVDF's piezoelectric properties under mechanical fatigue.

While PVDF is known for its fatigue resistance against crack propagation [10], understanding the mechanical fatigue resistance against PVDF's piezoelectric properties remains to be elucidated. In particular, the ferroelectric/piezoelectric properties of PVDF depend strongly on its polymorphs (α-, β-, γ- and δ-phases) [11,12]. Any fatigue-induced phase transformation, as well as mechanical failure, may cause decay in the material's piezoelectric properties [13,14]. Extensive research effort has been placed on controlling the phase content of PVDF, and various thermodynamic handles including temperature, stress or external electric field are known to affect the phase content. For instance, Sencadas et. al. reported that stretching PVDF induces α- to β-phase transformation [10]. To the best of our best knowledge, however, the phase evolution of poled PVDF under mechanical fatigue and its link to piezoelectric properties have not been studied extensively. Furthermore, no direct fatigue measurements of piezoelectric PVDF by applying repeated strain exists in the literature.

In this work, we assess the fatigue resistance of PVDF as piezoelectric harvesters by measuring the remnant polarization, piezoelectric voltage output, and relative phase contents after repeated mechanical input. To minimize the effects from electrode degradation, we employ commercially available Ag paste-based electrodes with protective sealing. This minimizes the change in the contact area between the electrode and PVDF layers during extended fatigue tests.

2. Materials and Methods

2.1. PVDF Fatigue Preparation

Figure 1 shows the unit PVDF capacitor (Piezo film sensor: DT1-028K, TE connectivity) used in this study and the stress application apparatus. The capacitor consists of 28 µm-thick PVDF layer and Ag paste-based electrodes on both sides (Figure 1a). The entire capacitor is wrapped with a protective polymer coating layer, ensuring that the paste-based electrodes stay intact during repeated bending. The dimensions of the unit capacitor are 16 mm wide, 41 mm long and 40 µm thick, with the measured capacitance of 1.38 nF. The unit capacitors are then attached to 230 µm-thick polyimide (PI) substrates on both sides (Figure 1b). Nickel mesh-based, 30 µm-thick double side tapes (Soluteta, Inc., Hwaseong, Korea) are used to attach the capacitors to the substrates, and the lead wires from the capacitors are protected with conducting carbon tapes to ensure tight attachment during repeated bending tests and minimize background noise.

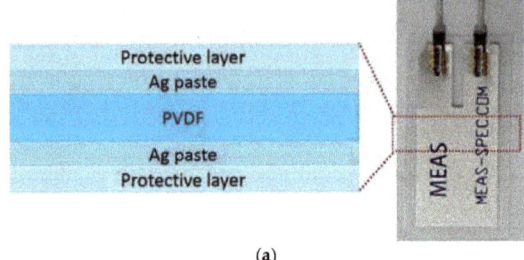

(a)

Figure 1. *Cont.*

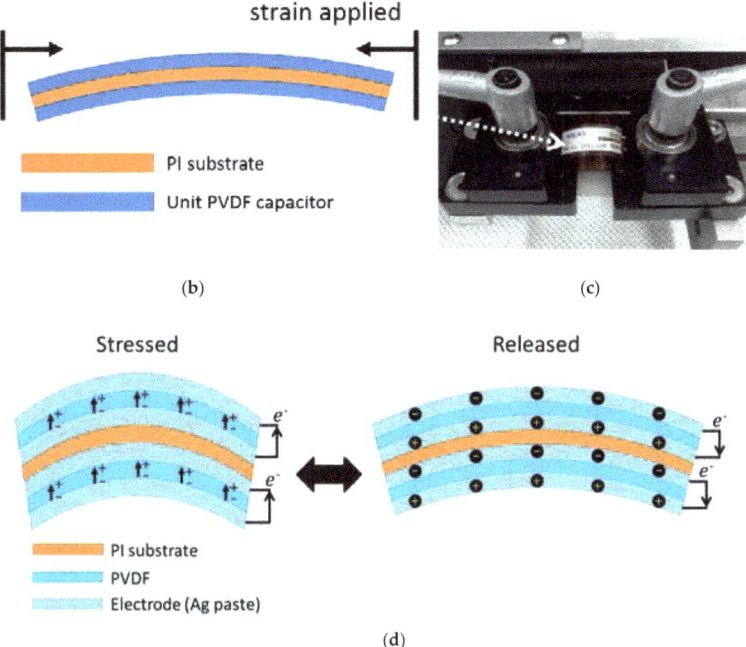

Figure 1. The test setup for piezoelectric PVDF's fatigue measurement system (**a**) Unit PVDF capacitor with silver paste electrode and sealing layer (**b**) PVDF based piezoelectric generator device with PI substrate (**c**) The homemade servo motor-based bending system with controlled frequency and input displacement. (**d**) The unit PVDF capacitor based bimorph-shaped device with dipole arrangement inside the PVDF and electric flow.

2.2. Fatigue Test Setup and Experimental Measurements

Figure 1c shows the linear motor set up to apply repeated strain to the device. The linear motor is based on a home-made servo motor with 420 rpm (revolution per minute). This is equivalent to 7 Hz bending at 13 mm horizontal displacement. As the motor moved inward, the mechanical stress is transmitted along the longitudinal direction of the unit capacitor. Bending the device generates tensile stress on the top capacitor and compressive stress on the bottom capacitor since the neutral axis lies at the center of the PI substrate. By tracking the change in radii of curvature, we estimate the amount of strain applied and the strain rate. The applied strain on both tensed and compressed PVDFs are 1.8% at the strain rate of 12.7%/s. The amount of strain is ensured to lie within the elastic limit (2.5%) of poled PVDF [15]. Figure 1d depicts the bimorph-shaped device with dipole arrangement inside PVDF and electric charge on the electrode. As we repeat bending and releasing the harvester using the home-made servo motor, it generates piezoelectric charges which can be extracted as an electric circuit.

2.3. Electrical Measurements–Voltage Output and Remnant Polarization

The open circuit voltage output is measured with a digital oscilloscope (DPO4014B, Tektronix, Seoul, Korea) during the entire course of bending. The bending is repeated up to 10^7 times, approximately 17 continuous days at 7 Hz. To accurately capture any decrease in the short-lived voltage peaks, we use the sampling rate of 500 pt/s.

After the completion of the fatigue tests, the Ag paste electrodes and the capping polymer layers are removed with ethanol. On the bare PVDF thin film, we deposit 100 nm-thick Pt electrodes via sputtering. The electrode area is maintained to be 1×1 cm^2 on both sides. The P-E hysteresis loops are measured by a ferroelectric testing system (RT66A, Radiant Technologies, Inc. Precision) connected with a high-voltage amplification interface (AMT-10B10, MATSUSADA Precision Inc., Kusatsu, Japan).

2.4. Phase Measurements by Raman Spectroscopy

The fatigued PVDF samples that remain free of Pt are examined with Raman spectroscopy to examine the relative phase content. The double side tape is used to attach the PVDF to the Si substrate. The PVDF thin films are analyzed by 532 nm green laser to observe PVDF Raman mode clearly. The Raman analysis is also carried out at room temperature to maintain the condition of fatigue tests. After 10^7 bending, we observed that the atomic ordering of PVDF β-phase is maintained.

3. Results and Discussions

Figure 2 shows the generated piezoelectric voltage for the tensed (Figure 2a–c) and the compressed (Figure 2d–f) PVDFs for the entire 10^7 cycles. In Figure 2b,e, we plot the averaged peak voltages for 60 peaks around the specified bending number (10^N) in logarithmic scale for better assessment of the fatigue behavior.

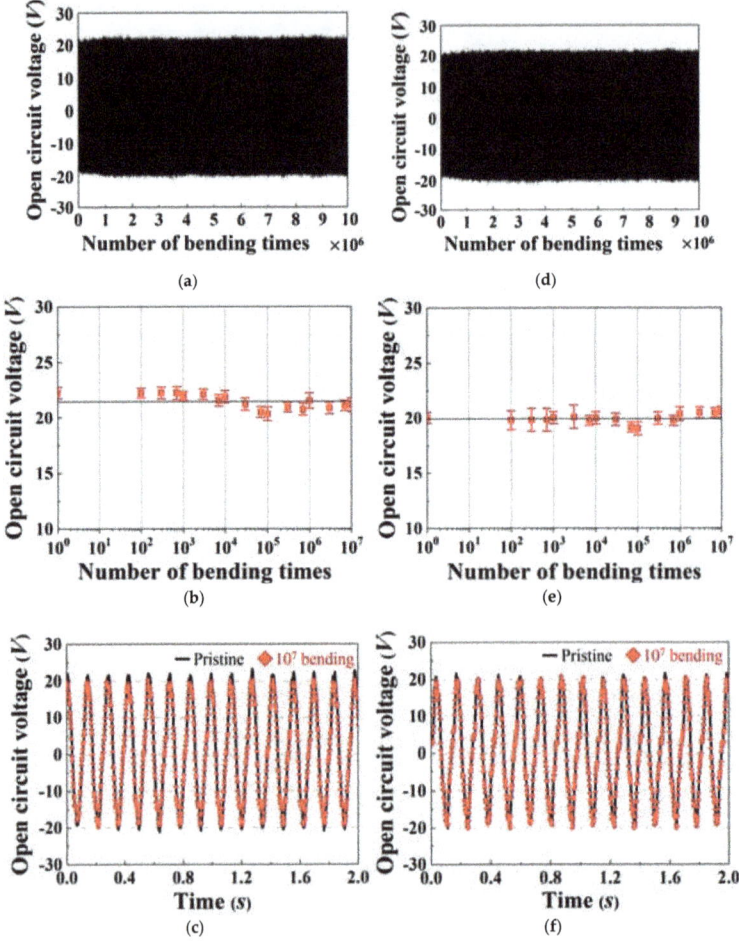

Figure 2. The measured open circuit voltages during repeated bending of 10^7 times (**a–c**) The open circuit voltage profile for tensed PVDF layers (**d–f**) those for compressed PVDF layers.

The relevant standard deviations are plotted as error bars in Figure 2b,e. For the tensed and compressed PVDF, the initial voltage output of 22.3 V and 20.0 V remain almost identical to 21.2 V and 20.5 V after 10^7 cycles, respectively. These amazingly stable voltage measures indicate that the piezoelectric property of PVDF does not readily change upon applying repeated mechanical tension or compression, given the stable electrode adhesion. The voltage response to 7 Hz mechanical input frequency does not exhibit any notable change in phase behavior or voltage peak width either, as shown in Figure 2c,f.

The polarization-electric field (PE) loops measured for the pristine and fatigued PVDF samples also reveal unaffected polarization properties of PVDF after repeated mechanical input. Figure 3 shows the polarization loops for the pristine, tensed and compressed PVDF. The remnant polarization of pristine PVDF is 5.48 µC/cm^2, similar to previously reported corona-poled PVDF in β-phase [16,17]. Since the piezoelectric constants (d_{ij}) are directly proportional to the remnant polarization values, assessing how fatigue affects the remnant polarization provides an indication of fatigue resistance against piezoelectricity loss.

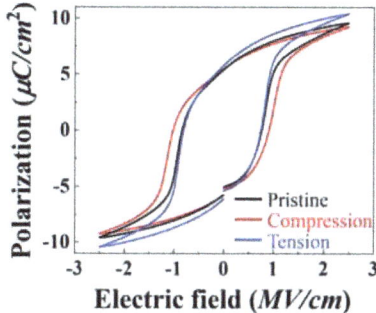

Figure 3. The PE hysteresis loop of PVDF at pristine and after applying 10^7 tensile and compressive stressed. The initial remnant polarization of 5.48 µm/cm^2 at pristine PVDF remains almost identical after tensile fatigue tests, however, decreases to 5.08 µm/cm^2 compressive fatigue tests.

The remnant polarization of PVDFs after 10^7 cycles of repeated tension and compression are 5.51 µC/cm^2 and 5.08 µC/cm^2, respectively. These numbers are clearly far higher than the polarization loss observed by Zhu et al. after repeated bias application [6]. Interestingly, the compressed sample exhibits a notable decrease of 7.3% in remnant polarization. The saturation polarization also decreases by 4.2% from 9.5 µC/cm^2 to 9.1 µC/cm^2 for the compressed sample, while that of the tensed sample increases by 9.4% to 10.4 µC/cm^2.

The anisotropic behavior of remnant polarization according to the stress direction calls for further investigation. While the decrease of 7.3% is not significant considering the excessive mechanical tests, the difference between tension and compression suggests that the fatigue resistance to de-poling may have an anisotropic response to the direction of stress. A potential explanation involves phase transformation into other polymorphs such as α- or γ-phase PVDF. Previous reports have shown that stretching a PVDF film results in α- to β-phase transformation [10]. While the strain is small in our fatigue tests compared to the stretch required for PVDF's phase transformations, fatigue-induced phase transformations with small strains have been reported in various materials systems [18]. To examine whether indeed repeated mechanical strain results in different phase behavior, we examine the phase content of the pristine, tensed and compressed PVDF layers with Raman spectroscopy.

Figure 4 plots the shift in Raman response for PVDF films. PVDF is known to possess phase-dependent characteristic Raman shifts: 839 cm^{-1} for the β-phase and 794 cm^{-1} for the α-phase [19]. We clearly observe the dominant β-phase in all three samples in Figure 4a, with notable peaks for the α-phase. Comparing the pristine sample with the compressed and tensed samples, we surprisingly observe increased α-phase content in tensed PVDF, and decreased α-phase content in compressed PVDF. Figure 4b plots the average intensity ratio (I_α/I_β) for four different pristine, compressed and tensed samples. The standard deviations are plotted as the error bars in the Figure 4b. Overall, the change is in the intensity is ratio is rather small after extensive mechanical tests. We also note that both the tensed and compressed samples exhibit increased error bar length compared to the pristine sample, with especially large deviation for the tensed sample. Although the results do not conclude that a clear anisotropic response to stress direction during mechanical fatigue exists, the increased deviation calls for further investigation into fatigue-induced phase transformation in PVDFs. Tracking the absolute amounts of the α-phase, β-phase and the non-crystalline phases may help us further clarify the effect of repeated mechanical stress on PVDF's phase behavior. However, experimental difficulty exists for the current setup.

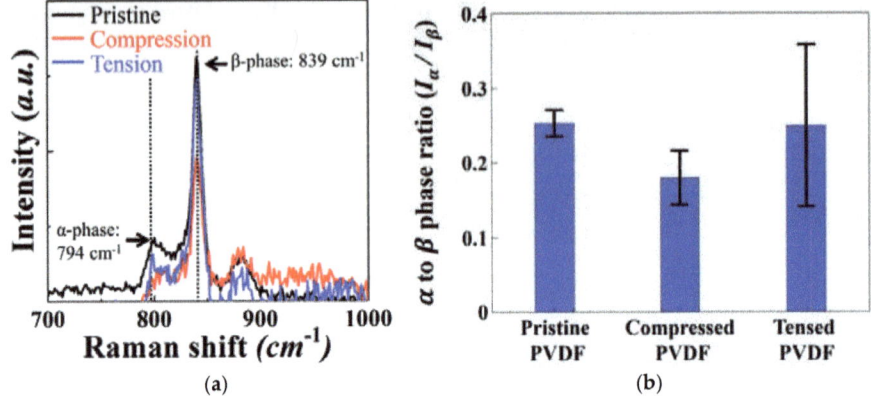

Figure 4. (a) The Raman shift observed for PVDF at pristine and after applying 10^7 tensile and compressive stress. (b) The average intensity ratio (I_α/I_β) with the standard deviation plotted as the error bars.

The stable piezoelectric properties of PVDF, demonstrated in this work by stable voltage output, remnant polarization and relative phase contents, not only suggests that PVDF possesses strong fatigue resistance against mechanical integrity loss or de-poling via domain switching. These results confirm the reports by Zhao et al. that the previously measured polarization loss during polarization fatigue likely results from the electrode delamination [7]. Our results also show that the electrode delamination may be easily engineered via using soft metal-based paste capped with the suitable sealing layer. As the material exhibits wide applicability ranging from piezoelectric generators or actuators to ferroelectric capacitors, its fundamental fatigue resistance calls for scientific understanding. Su et al. reported through first principles based calculations that nucleating nonpolar kinks in the PVDF chain (equivalent to seed for nonpolar phases) requires 24.8 kcal/mol [15]. In addition, the mobility of the polar-nonpolar interfaces requires the threshold tensile stress of approximately 2 GPa, a very high stress for soft PVDF with Young's modulus of 2.9 GPa [20].

4. Conclusions

In summary, we report the excellent mechanical fatigue resistance of piezoelectric PVDF during 10^7 cycles of mechanical loading. PVDF exhibits strong mechanical fatigue resistance against both tension and compression. The stable voltage output and remnant polarization suggest both little mechanical degradation such as cracking or pore formation and de-poling behavior such as domain switching. A 7% decrease in remnant polarization under repeated compression suggests a potential anisotropic response to stress directions. The repeated mechanical tests also result in deviations in phase behavior. The works demonstrate that PVDF-based piezoelectric energy harvesters or sensors possess suitable fatigue resistance for extended-life applications.

Author Contributions: C.-Y.K. and S.K. conceived the project. Y.-H.S. and I.J. performed the fatigue tests and H.P and S.K. performed the polarization-electric field loop measurements. J.J.P. and Y.-H.S. performed the Raman spectroscopy measurements. All authors participated in discussion and analyses of the data. S.K., Y.-H.S. and C.-Y.K. wrote the initial manuscript and all authors participated in editing and proofreading the manuscript.

Acknowledgments: The work was supported by Energy Technology Development Project (KETEP) grant funded by the Ministry of Trade, Industry and Energy, Republic of Korea (Piezoelectric Energy Harvester Development and Demonstration for Scavenging Energy from the Road Traffic System, Project no. 20142020103970) and the KU-KIST Research Program of Korea University (R1309521). S.K. and C.-Y.K. acknowledge support from support from the National Research Council of Science & Technology (NST) grant by the Korea government (MSIP) (No. CAP-17-04-KRISS).

Conflicts of Interest: The authors declare no conflict of interest.

References

1. Jung, I.; Shin, Y.-H.; Kim, S.; Choi, J.; Kang, C.-Y. Flexible piezoelectric polymer-based energy harvesting system for roadway applications. *Appl. Energy* **2017**, *197*, 222–229. [CrossRef]
2. Lee, C.S.; Kim, J.Y.; Lee, D.E.; Joo, J.; Han, S.; Beag, Y.W.; Koh, S.K. An approach to durable poly(vinylidene fluoride) thin film loudspeaker. *J. Mater. Res.* **2003**, *18*, 2904–2911. [CrossRef]
3. Yu, Y.; Sun, H.; Orbay, H.; Chen, F.; England, C.G.; Cai, W.; Wang, X. Biocompatibility and in vivo operation of implantable mesoporous PVDF-based nanogenerators. *Nano Energy* **2016**, *27*, 275–281. [CrossRef] [PubMed]
4. Xiong, H.; Wang, L. Piezoelectric energy harvester for public roadway: On-site installation and evaluation. *Appl. Energy* **2016**, *174*, 101–107. [CrossRef]
5. Shin, Y.-H.; Jung, I.; Noh, M.-S.; Kim, J.H.; Choi, J.-Y.; Kim, S.; Kang, C.-Y. Piezoelectric polymer-based roadway energy harvesting via displacement amplification module. *Appl. Energy* **2018**, *216*, 741–750. [CrossRef]
6. Zhu, G.; Zeng, Z.; Zhang, L.; Yan, X. Polarization fatigue in ferroelectric vinylidene fluoride and trifluoroethylene copolymer films. *Appl. Phys. Lett.* **2006**, *89*, 102905. [CrossRef]
7. Zhao, D.; Katsouras, I.; Li, M.; Asadi, K.; Tsurumi, J.; Glasser, G.; Takeya, J.; Blom, P.W.M.; de Leeuw, D.M. Polarization fatigue of organic ferroelectric capacitors. *Sci. Rep.* **2015**, *4*, 5075. [CrossRef] [PubMed]
8. Sim, G.-D.; Hwangbo, Y.; Kim, H.-H.; Lee, S.-B.; Vlassak, J.J. Fatigue of polymer-supported Ag thin films. *Scr. Mater.* **2012**, *66*, 915–918. [CrossRef]
9. Sim, G.-D.; Won, S.; Lee, S.-B. Tensile and fatigue behaviors of printed Ag thin films on flexible substrates. *Appl. Phys. Lett.* **2012**, *101*, 191907. [CrossRef]
10. Sencadas, V.; Gregorio, R.; Lanceros-Méndez, S. α to β Phase Transformation and Microstructural Changes of PVDF Films Induced by Uniaxial Stretch. *J. Macromol. Sci. Part B* **2009**, *48*, 514–525. [CrossRef]
11. Gregorio, R. Determination of the α, β, and γ crystalline phases of poly(vinylidene fluoride) films prepared at different conditions. *J. Appl. Polym. Sci.* **2006**, *100*, 3272–3279. [CrossRef]
12. Salimi, A.; Yousefi, A.A. Analysis Method. *Polym. Test.* **2003**, *22*, 699–704. [CrossRef]
13. Okayasu, M.; Odagiri, N.; Mizuno, M. Damage characteristics of lead zirconate titanate piezoelectric ceramic during cyclic loading. *Int. J. Fatigue* **2009**, *31*, 1434–1441. [CrossRef]
14. Wang, D.; Fotinich, Y.; Carman, G.P. Influence of temperature on the electromechanical and fatigue behavior of piezoelectric ceramics. *J. Appl. Phys.* **1998**, *83*, 5342–5350. [CrossRef]
15. Tavares, C.J.; Marques, S.M.; Rebouta, L.; Lanceros-Méndez, S.; Sencadas, V.; Costa, C.M.; Alves, E.; Fernandes, A.J. PVD-Grown photocatalytic TiO_2 thin films on PVDF substrates for sensors and actuators applications. *Thin Solid Films* **2008**, *517*, 1161–1166. [CrossRef]
16. Guy, I.L.; Limbong, A.; Zheng, Z.; Das-Gupta, D.K. Polarization fatigue in ferroelectric polymers. *IEEE Trans. Dielectr. Electr. Insul.* **2000**, *7*, 489–492. [CrossRef]
17. Katsouras, I.; Asadi, K.; Li, M.; van Driel, T.B.; Kjær, K.S.; Zhao, D.; Lenz, T.; Gu, Y.; Blom, P.W.M.; Damjanovic, D.; et al. The negative piezoelectric effect of the ferroelectric polymer poly(vinylidene fluoride). *Nat. Mater.* **2016**, *15*, 78–84. [CrossRef] [PubMed]

18. Wang, C.-C.; Mao, Y.-W.; Shan, Z.-W.; Dao, M.; Li, J.; Sun, J.; Ma, E.; Suresh, S. Real-time, high-resolution study of nanocrystallization and fatigue cracking in a cyclically strained metallic glass. *Proc. Natl. Acad. Sci. USA* **2013**, *110*, 19725–19730. [CrossRef] [PubMed]
19. Riosbaas, M.T.; Loh, K.J.; O'Bryan, G.; Loyola, B.R. In situ phase change characterization of PVDF thin films using Raman spectroscopy. In *Sensors and Smart Structures Technologies for Civil, Mechanical, and Aerospace Systems 2014*; Lynch, J.P., Wang, K.-W., Sohn, H., Eds.; SPIE Digital Library: San Diego, CA, USA, 2014; p. 90610Z.
20. Su, H.; Strachan, A.; Goddard, W.A. Density functional theory and molecular dynamics studies of the energetics and kinetics of electroactive polymers: PVDF and P(VDF-TrFE). *Phys. Rev. B* **2004**, *70*, 064101. [CrossRef]

© 2018 by the authors. Licensee MDPI, Basel, Switzerland. This article is an open access article distributed under the terms and conditions of the Creative Commons Attribution (CC BY) license (http://creativecommons.org/licenses/by/4.0/).

Review

The Progress of PVDF as a Functional Material for Triboelectric Nanogenerators and Self-Powered Sensors

Jin Pyo Lee, Jae Won Lee and Jeong Min Baik *

School of Materials Science and Engineering, KIST-UNIST-Ulsan Center for Convergent Materials, Ulsan National Institute of Science and Technology (UNIST), Ulsan 44919, Korea; leejp@unist.ac.kr (J.P.L.); leejw@unist.ac.kr (J.W.L.)
* Correspondence: jbaik@unist.ac.kr; Tel.: +82-052-217-2324

Received: 13 September 2018; Accepted: 16 October 2018; Published: 20 October 2018

Abstract: Ever since a new energy harvesting technology, known as a triboelectric nanogenerator (TENG), was reported in 2012, the rapid development of device fabrication techniques and mechanical system designs have considerably made the instantaneous output power increase up to several tens of mW/cm^2. With this innovative technology, a lot of researchers experimentally demonstrated that various portable/wearable devices could be operated without any external power. This article provides a comprehensive review of polyvinylidene fluoride (PVDF)-based polymers as effective dielectrics in TENGs for further increase of the output power to speed up commercialization of the TENGs, as well as the fundamental issues regarding the materials. In the end, we will also review PVDF-based sensors based on the triboelectric and piezoelectric effects of the PVDF polymers.

Keywords: PVDF; sensor; triboelectric nanogenerator

1. Introduction

1.1. Background of Energy Harvesting Based on the Triboelectric Effect

Most recently, a newly designed energy generating device named the triboelectric nanogenerator (TENG) was reported and various types of TENGs have been demonstrated, proven as a highly efficient, simple, robust and cost-effective technique for efficiently converting various mechanical energies around us to electricity [1–18]. In principle, the electrical energy is generated as two different materials are brought into contact with each other, in conjunction with the triboelectric electrification and electrostatic induction. The sources of mechanical energy such as winds and moving things are available anywhere and anytime in our surroundings [18,19], thus, TENG will be appropriate as a power-supply unit for portable devices although the energy is not big as expected [20]. Since the first demonstration of the TENG on 2012, various TENG structures and new functional materials brought about the significant increase of the instantaneous power density up to several tens of mW/cm^2, as shown in Figure 1 and Table 1 [1,2,12,21–29]. The output current also increased up to several mA although it seemed that there was no significant increase recently. By virtue of the increased output power, it was successfully demonstrated that it was possible to power up various portable electronic devices as well as to fabricate the self-powered physical and chemical sensors. Actually, many portable/wearable devices such as smart phones/watches, healthcare sensors, smart goggles, etc., could be charged by the TENGs, which generated enough power to turn on or use the devices for a moment. The TENGs also made various sensors which could mechanically detect the ambient pressure [16,30–33], motion [34], airflow [35], vibration [36], and chemically detect various chemical/molecular species including methanol [37], ethanol [38–40], mercury [41], glucose [42], phenol [43], metal ions [44] without any external power.

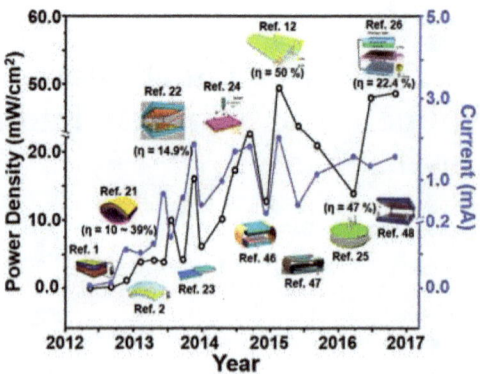

Figure 1. The areal power densities and output currents of various triboelectric nanogenerators (TENGs) reported for the last 5 years [45].

Table 1. The areal power densities, output currents, and charge densities of various triboelectric nanogenerators (TENGs) reported for the last 5 years.

	Positively Charged Materials	Negatively Charged Materials	Working Mode	Output Power (mW/cm²)	Charge Density (μC/m²)
#1 [1]	PET	Kapton	Vertical-contact	0.00036	-
#2 [2]	PET	PDMS	Vertical-contact	0.00234	-
#3 [21]	Al	PDMS	Vertical-contact	3.56	-
#4 [22]	Au	PDMS	Vertical-contact	31.3	594.2 (calculated)
#5 [23]	Nylon	PTFE	Lateral-sliding	0.53	59
#6 [24]	Al	FEP	Vertical-contact	31.5	240
#7 [46]	Al	PVDF	Vertical-contact	0.26	360.2
#8 [12]	Cu	PTFE	Freestanding	50	323
#9 [47]	Al	ZnSnO3-PVDF (composites)	Vertical-contact	3	101.3
#10 [25]	Cu	Kapton	Lateral-sliding	13.2	-
#11 [26]	Al	PDMS	Vertical-contact	46.8	270
#12 [48]	Al	PVDF-Gn	Vertical-contact	2.6	23

1.2. The Four Fundamental Modes of the Triboelectric Nanogenerator

Up to now, there are two representative modes in TENGs such as the vertical contact-separation mode and lateral sliding mode depending on the direction. An object moves with respect to the other one. In the vertical contact-separation mode, the working mechanism of the TENG is understood as the potential difference occurring due to the contact and separation of two surfaces, as shown in Figure 2a. When an external force is applied onto the top surface of the device, the surface comes into contact with the other surface and the charge transfer occurs between them, generating positive charges on the surface of one material and negative charges on the other one. The charge transfer is mainly determined by the intrinsic properties of the materials, known as the triboelectric series. As the force is released, the two surfaces with opposite charges are separated, inducing a potential difference across the top and bottom electrodes. This is the driving force of the electron flow through the external circuit from one electrode to the other one.

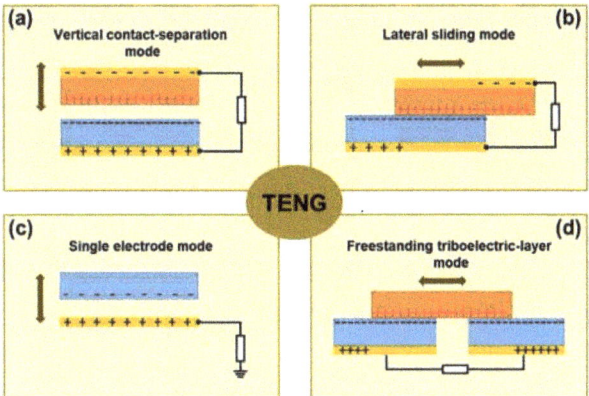

Figure 2. The four fundamental operation modes of the triboelectric nanogenerators. (**a**) The vertical contact-separation mode. (**b**) The lateral sliding mode. (**c**) The single-electrode mode. (**d**) The free-standing mode.

In the second operation mode, the two surfaces are brought into contact in parallel with each other, named as the lateral sliding mode, as shown in Figure 2b. Once the top surface with positive charges slides outward, the decrease of the contact area induces the flow of electrons through the external circuit because of the in-plane charge separation generating an electric field parallel to the plates. Subsequently, when the top surface reverts to the original position, the electrons on the top electrode will flow back through the external circuit to the bottom electrode in the same manner. Compared with the vertical contact-separation mode, the lateral sliding mode shows a quite high energy conversion efficiency and is quite promising in wearable application because of the low input energy and various sliding motions occurring during human movement.

Additionally, there were two more operation modes, the single electrode mode, and free-standing mode. Specifically, the single electrode mode has only one electrode, as shown in Figure 2c. In the mode, the surfaces of the polymer and metal electrode are brought into contact with each other and the charge transfer occurs between them. Once the polymer and the electrode are separated, the negative charges on the surface of the polymer induced positive charges on the electrode, driving free electrons to flow from the electrode to the ground, creating a current. The free-standing mode triboelectric nanogenerator can scavenge electrical energy without an attached electrode, as shown in Figure 2d. The TENG consists of two electrodes underneath a dielectric layer and the other frictional layer. The frictional layer approaching to and departing from the electrodes creates an unbalanced charge distribution, causing the electrons to flow between the two electrodes to balance the local potential distribution. It is well-known that this mode could generate quite a high energy by the optimization of the electrode pattern and the proper choice of the materials.

1.3. The Importance of Dielectrics in the Triboelectric Nanogenerator

As mentioned above, the instantaneous areal power density has reached up to several tens of mW/cm^2. However, this does not mean the improvement of the energy conversion efficiency. It tells us facts that the fabricated TENGs can generate the output powers at confined input conditions (i.e., large input force or optimized input conditions). The areal output currents reported are also still less than 1 mA/cm^2, too low to provide enough energy for electronic devices. Recently, it seems that there was no more significant increase in the output power of the TENGs.

It is well-known that the output current of the TENGs is determined by the density of the surface charges transferred between the two materials. As a positively charged material, aluminum (Al) was widely used because of the excellent electron donor characteristics although there have been promising candidates consisting of Au, Cu, etc. On the other hand, various polymers such as polydimethylsiloxane (PDMS) [49,50], polytetrafluoroethylene (PTFE) [51–53], fluorinated ethylene propylene (FEP) [54,55], and polyimide (PI) [56–58] have been tested as negatively charged materials. Among many polymers, polyvinylidene difluoride (PVDF) has many advantages such as a large dipole moment [59], good formability [60], and flexibility [61], thus, it has been considered as a good candidate for an effective dielectric in TENG. Until now, various key approaches have been applied to develop promising materials for the enhancement of the output performance, such as the work function, the dielectric constant, the functional group (−F), the frictional coefficient, the surface resistivity, the carrier density, the intrinsic carrier density, etc. Among the approaches, a high dielectric constant and high work function were mainly studied theoretically and mathematically. In virtue of the dielectric constant, the total transferred charge density (σ') in the metal-polymer system can be expressed by [62]

$$\sigma' = \frac{-\sigma_0 d_{gap}}{d_{gap} + \frac{d_{dielectric}}{\varepsilon_{dielectric}}} \quad (1)$$

where σ_0 is the triboelectric surface charge density at the equilibrium state, d_{gap} and $d_{dielectric}$ are the gap distance and the thickness of the dielectric films, and $\varepsilon_{dielectric}$ is a dielectric constant of the dielectric films. Thus, the charge density will be increased by the increase of the dielectric constant. The charge density is also influenced by the surface chemical potential difference between the two materials, according to the following general equation [27]

$$\sigma = \frac{[(W - E_0)e]\left(1 + \frac{t}{\varepsilon z}\right)}{\frac{t}{\varepsilon \varepsilon_0} + \left(1/N_s(E)e^2\right)\left(1 + \frac{t}{\varepsilon z}\right)} \quad (2)$$

where $W - E_0$ is the difference in the effective work functions between the two materials, e, t, ε, ε_0, z, and $N_s(E)$ are the charge of an electron, distance of space, relative permittivity of the dielectric, vacuum permittivity of free space, thickness of the dielectric film, and the averaged surface density of states. This indicates that as the work function increases, the surface charge density accordingly increases. However, despite the above-detailed studies, the charge density by the artificial charge injection is practically limited and easily lost at ambient conditions. It implies that dielectric materials that can efficiently accept many charges by physical contact need to be developed.

2. Polyvinylidene Fluoride (PVDF) as an Effective Dielectric in Triboelectric Nanogenerators

PVDF, a highly non-reactive polymer with a strong piezoelectric response and low acoustic impedance, has been regarded as one of the most popular polymers in energy harvesting technologies. It also shows a good flexible and mechanical properties, making it a good candidate as an effective dielectric in triboelectric nanogenerator. It is a linear homopolymer produced by the repetition of CH_2-CF_2 monomers, crystallized into four possible conformations named as the α, β, γ, and δ phases. In general, they possess a permanent dipole moment. When the dipoles are aligned in one direction via mechanical stretching or an electrical poling process under a high electric field, the highest dipole moment can be obtained and this type of structure is known as β-phase PVDF film. The β-phase PVDF structure of PVDF has fluorine atoms and hydrogen atoms in the opposite direction, forming a high dipole density. Thus, it intrinsically has a high dielectric constant (~10), compared to other polymers. In order to additionally the dielectric constant, various ferroelectric particles such as $BaTiO_3$ [63], $ZnSnO_3$ [47], and $SrTiO_3$ [64] are blended and functional groups such as -TrFE [28], -HFP [29], and -PtBA [48] are grafted via copolymerization technique. These fluorine functional groups exhibit large electron affinity, increasing the charges-accepting characteristics from the metal during the instantaneous contact. With this material synthesis by exploiting intrinsic properties of ferroelectric materials, new dielectric materials should be developed to increase the energy conversion efficiency for sustainable self-powered devices.

2.1. PVDF-Based Polymers as an Effective Dielectric

The Wang group, for the first time, used PVDF as a dielectric without any modifications [46]. As shown in Figure 3a, the TENG consists of two layers, a nanoporous aluminum (Al) foil and a PVDF thin film deposited on a copper substrate with Kapton films of a thickness of 125 μm as a spacer. During the cycled contact under a compressive force of around 50 N, the TENG produced a short-circuit current density (J_{sc}) of 6.13 μA/cm² (Figure 3b). After forward and reverse polarization, the J_{sc} increased to 8.34 μA/cm² and 4.83 μA/cm², respectively. By averaging the instantaneous peaks, the forward-polarized TENG showed an output performance 3 times larger than that of the reverse-polarized TENG. They explained the enhancement in terms of the surface potential of the PVDF thin film modified by the polarization process, as shown in Figure 3c. When the film was forward-polarized, the net dipole moment was increased because of the arrangement of the dipoles along the same direction. This reduced the characteristic energy level of the PVDF, thereby increasing the potential difference with the Fermi level of Al. This resulted in an enhancement of the charge-transfer characteristics between the PVDF and the Al electrode. The light-emitting diode (LED) connected to the forward-polarized TENG was much brighter than that of the reverse-polarized TENG (Figure 3d). This shows the experimental evidence of the relationship between the work function difference between the PVDF and Al, and the electric outputs of the TENG.

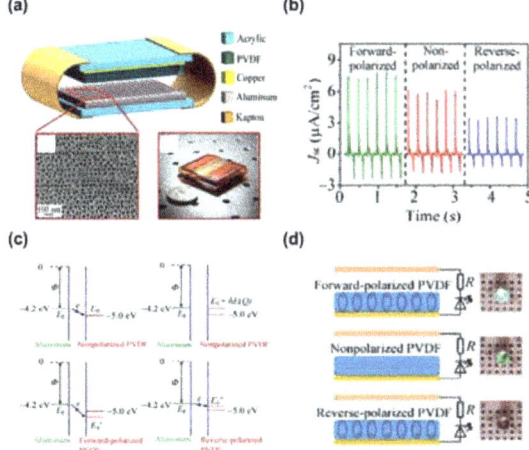

Figure 3. (a) The schematic of the TENG with a double-layer structure. A SEM image of nanopores on an aluminum foil. Photograph of a TENG. (b) J_{sc} of TENGs fabricated using different types of polyvinylidene fluoride (PVDF) films under a periodic compressive force of around 50 N applied by an electric shaker at a frequency of 4 Hz. (c) Schematic energy band diagram illustrating the process when electrons tunnel between E_F and E_0. The electric field developed by charge transfer raises the energy of the electrons in the PVDF's surface states by $\delta E(Q)$. The positive bond charges of the forward-polarized PVDF reduce the characteristic energy level of PVDF to $E_{0'}$. The negative bond charges of the reverse-polarized PVDF increase the characteristic energy level of PVDF to E_0''. (d) Circuit diagrams and the different performances among TENGs fabricated using different types of PVDF when they were used as direct power source for LED bulbs. Reproduced with permission from Wang et al., *Nano Research* **2015**, *15*, 990–997. Copyright 2015 Springer Nature.

The PVDF nanofibers of approximately 790 nm in diameter by the electrospinning method were also prepared, deposited on Kapton film, and contacted with the nylon nanofibers as shown in Figure 4a [65]. Based on the triboelectric series, negative charges are transferred to the PVDF from the nylon, creating instantaneous electrons flow through the external circuit between two electrodes. The TENG produced an instantaneous output power of 26.6 W/m² and it was demonstrated that the nanofibers were so effective in increasing the output current of TENGs, compared with the films as shown in Figure 4b. Thus, electrospinning, as a fiber production method, is appropriate for fabricating roughened surfaces to enhance the output power. The nanofibers also seem so flexible, quite helpful for harvesting wearable energy sources. The TENG was also connected to Zener diodes and a ZnO nanowire sensor to demonstrate the possibility of the TENG as a power source for a self-powered UV sensor, which was successfully proven (Figure 4c).

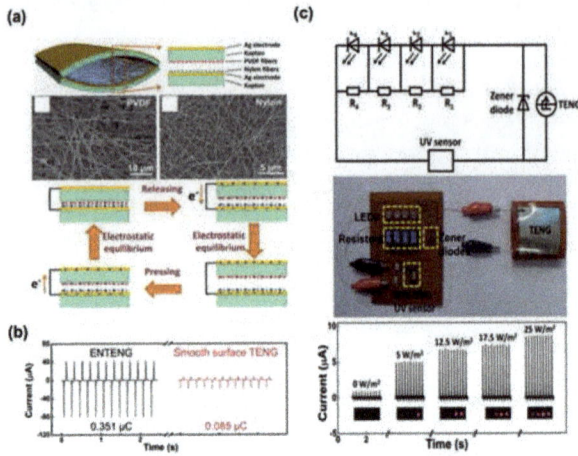

Figure 4. (a) The schematic of the TENG based on electrospun nanowires (ENTENG). SEM images of the electrospun PVDF and nylon nanofibers. The charge distribution in the device and the flowing direction in the circuit when the device is pressed and released. (b) The output current of the ENTENG and the output current of the smooth surface TENG with a similar device structure. (c) The schematic and optical images of the ultraviolet radiation (UVR) level detection system powered by an ENTENG. Current flowing across the UV sensor and the optical images of the LEDs at UV intensities of 0, 5, 12.5, 17.5, and 25 Wm^{-2}. Reproduced with permission from Jing et al., *Nanoscale* **2014**, *6*, 7842–7846. Copyright 2014 The Royal Society of Chemistry.

Very recently, there has been some progress in developing PVDF-based polymers as an effective dielectric in TENG. Baik group successfully synthesized poly (tert-butyl acrylate) (PtBA)-grafted PVDF copolymers through an atom transfer radical polymerization technique and demonstrated that the copolymers were very effective in increasing output performance, compared with pristine PVDF [48]. Unlike the PVDF copolymers such as P(VDF-TrFE), P(VDF-HFP) was mainly composed of β-phase. The copolymer was reported to be mostly composed of α phases with enhanced dipole moments by π-bonding and polar characteristics of the ester functional groups in the PtBA, bringing about the increase of the dielectric constant by approximately twice. In order to understand the effect of the dielectric constant on the output signal of the TENG, a very flat PVDF surface was made by the peeling-off process, as shown in Figure 5a. As the grafting ratio increased to 18%, the dielectric constant was increased from 8.6 to 16.5 in the frequency range of 10^2–10^5 Hz. This was attributed to the increase of the net dipole moment, as shown in Figure 5b. This increase in the dielectric constant enhanced the charge density that can be accumulated on the surface, generating the output voltage and current density of 64.4 V and 18.9 µA/cm^2, respectively, twice enhancement in both, compared to pristine PVDF based nanogenerator as shown in Figure 5c. To prove the results, they measured the accumulated surface charges density with different dielectric constant values, which showed an excellent agreement with the measured output current values as shown in Figure 5d. This paper, for the first time, experimentally proved that the output power linearly increased with the dielectric constant. However, the output power was still low because of the low dielectric constant value.

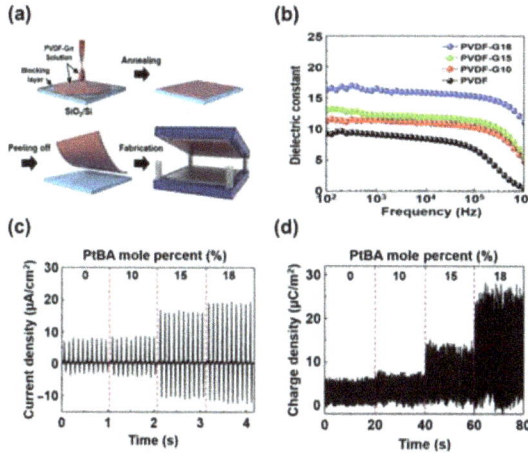

Figure 5. (a) The schematic diagrams of the fabrication process for the PVDF-Gn–based TENGs. (b) Frequency dependence of the dielectric constant values with various PtBA mole percentages ranging from 0 to 18%. (c) Output current densities generated by the PVDF-based TENGs as a function of the PtBA mole percentage ranging from 0 to 18%. (d) Charge densities generated by the PVDF-based TENGs as a function of the PtBA mole percentage ranging from 0 to 18%. Reproduced with permission from Baik et al., *Sci. Adv.* **2017**, *3*, 1500661. Copyright 2017 American Association for the Advancement of Science.

2.2. PVDF-Based Polymers Hybridized with Inorganic Materials

As mentioned above, despite the many advantages of PVDF polymers as a dielectric, it still had a low dielectric constant, thereby it limits the output power of the TENG. Thus, PVDF polymers embedded with inorganic nanoparticles have been utilized as dielectrics in TENGs. Zinc stannate ($ZnSnO_3$) materials with a strong piezoelectric response and huge spontaneous polarization value have been widely used as a candidate in polymers—such as PDMS—in which most of those studies have been focused on piezoelectricity or the piezoelectric nanogenerator. Recently, a TENG with the vertical contact-separation mode which consisted of $ZnSnO_3$ nanocubes-PVDF composites and a polyamide-6 (PA6) membrane was reported, as shown in Figure 6a. The high dispersion of $ZnSnO_3$ in the composites was obtained by increasing the shear speeds during the compounding process. By the phase inversion process, the composites showed that a porous structure only exists on the surface in which the pore size is in the range of 0.5–1.0 µm, as shown in Figure 6b. Compared with the pristine PVDF, the composites also showed the increase of the dielectric permittivity corresponding to a higher polarization and a higher piezoelectric coefficient d_{33} of −65 pm/V. In the TENG, the PA6 membrane was attached by using an adhesive aluminum tape electrode on one side and a PVDF or PVDF-$ZnSnO_3$ composite attached to the opposite side using aluminum tape. The composites are contacted with a PA6 membrane, showing an increase of the instantaneous output signals such as 520 V and 2.7 mA/m^2 in output voltage and output current, respectively. It corresponds to an enhancement of 70% and 200%, respectively as shown in Figure 6c.

In addition to the inorganic nanoparticles, PVDF based nanocomposites composed of PVDF-PBA-MWCNTs (PCNTs) were successfully synthesized, as shown in Figure 7a [66]. With the incorporation of PCNTs into the PVDF matrix, the crystalline phase transition significantly occurred from the α-phase to the β-phase, meaning that the β-phase was enhanced. It was explained in terms of the enhancement of compatibility of the PCNTs with CF_2 groups in the PVDF by the carboxylic acid groups, increasing the nucleation sites for the β-phase. Under the mechanical deformation at 3 Hz, the PVDF-10PCNT composite showed an output performance enhancement of almost 8 times compared to the pristine PVDF film by the physical contact with the Al film, as shown in Figure 7b.

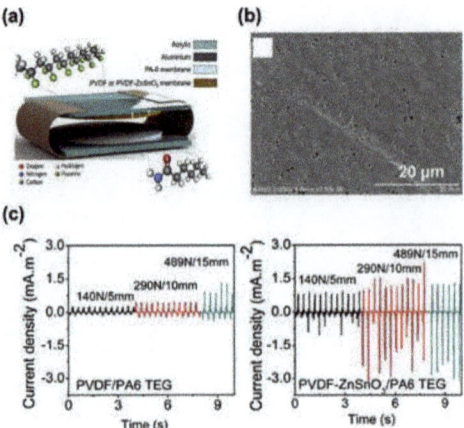

Figure 6. (a) The structure of the triboelectric nanogenerator (TEG) showing the position of the various components. (b) Scanning electron microscopy images of the surface and thickness cross-section of the PVDF-ZnSnO$_3$ membrane. (c) Short circuit current values for PVDF/PA6 TEG and PVDF-ZnSnO$_3$/PA6 TEG. Reproduced with permission from Luo et al., *Nano Energy* **2016**, *30*, 470–480. Copyright 2016 Elsevier.

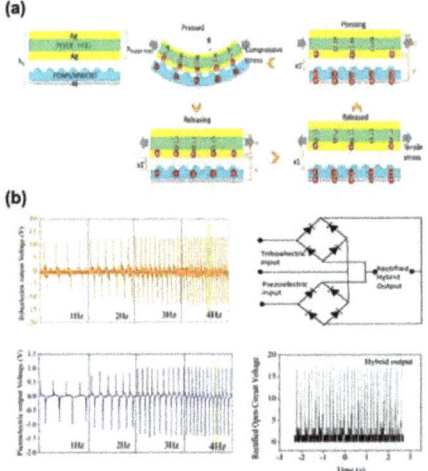

Figure 7. (a) The cross-section structural schematic; Charge distribution of triboelectric and piezoelectric effect during force pressing and releasing. Inside of the picture, $0 < x_1' < x_1$, h_s is the summation height of TPENG, $h_{P\,(VDF\text{-}TrFE)}$ is the height of P(VDF-TrFE) film, R is the stretching radius. (b) The open-circuit performance of the triboelectric and piezoelectric under the force of 5 N as a function of frequency. The circuit diagram of the hybrid output in which the piezoelectric and triboelectric outputs are combined in parallel. Hybrid open-circuit output voltage. Reproduced with permission from Luo et al., *Sci. Rep.* **2016**, *6*, 36409. Copyright 2016 Springer Nature.

3. PVDF Based Applications

In addition to power generation, PVDFs have been also extensively studied in sensor applications by many researchers. Actually, PVDFs have attracted much attention as an important material in various applications for its exceptional properties such as its high thermal stability, outstanding chemical resistance, low acoustic impedances, low permitivities, flexibility, and membrane forming properties. Additionally, PVDF is pyroelectric, meaning that it has a higher temperature dependent performance compared to other ceramic based sensors. With these excellent properties of PVDF, human skin-inspired multimodal e-skins were developed by mimicking various structures and functions of the elaborate sensory system in human finger-tips [67]. Here, polymer composites consisted of PVDF and reduced graphene oxide (rGO) were synthesized and a micro-structured ferroelectric skin was fabricated by mimicking the sensor system in human fingertips, in which the piezoresistive change between interlocked microdome arrays of rGO/PVDF composites was measured. This e-skin was also multifunctional and could detect pressure, vibration, and temperature with quite high sensitivities. A free-standing ZnO nanorods/PVDF composites are synthesized to fabricate a multifunctional tactile sensor that can independently measure the pressure and temperature [68]. Due to the enhanced β-phase of the PVDF film by the ZnO nanostructures, the minimum detectable pressure of the sensor was about 10 Pa and could detect temperatures in the range of 20–120 °C.

However, most of the research on the sensor based PVDF relies on the piezoelectric/ferroelectric properties of the material. Recently, there have been some reports on the sensing performance based on triboelectric effects in which the PVDF-based materials were used as positively charged layers to utilize its piezoelectric properties, thus, to fabricate the hybrid nanogenerator [69]. To enhance the flexibility and stretchability, and thus, the contact uniformity, the PVDF nanofibers were fabricated by the electrospinning technique. The hybridization was very effective in increasing the output voltage when both outputs were combined in parallel. It was also reported that it could exhibit a high sensitivity of 0.068 V/kPa in the range of 100 and 700 kPa.

4. Conclusions

Herein, various TENG technologies including the device structures, operation modes, a few functional materials, as well as the fundamental issues such as the working mechanism, and the contribution to the substantial increase of the output power up to several tens of mW/cm^2, were reported. We also reviewed PVDF-based materials as a promising candidate for efficient dielectrics in TENGs with a high output performance. PVDFs have many advantages such as their excellent mechanical properties (e.g., flexibility) and molecular structures showing high ferroelectric properties and high electron affinities due to the C-F bonds, making it possible to synthesize new dielectrics for further increases in the output power of TENGs. Up to now, although there were a few, not many reports on the synthesis of the PVDF-based materials exist as most of the papers have focused on the increase of the dielectric constant via the copolymerization techniques or the hybridization with inorganic nanomaterials. This strategy, together with the increase of surface roughness, was proven to be very effective in increasing the output power of the TENG. However, despite many efforts, the output power may not be enough for the realization of self-powered systems. From the material's aspect, the efficiency of the charge transfer to the dielectric and the capability of holding the charge in the dielectric need to be improved. The fabrication of TENGs based on the multilayered dielectric film has been reported as being quite effective in increasing the output power. Thus, it is thought that the TENGs will be used as portable power sources for low-power electronic devices such as sensors because of the low outputs. The TENG can also be utilized to fabricate physical sensors which are capable of sensing strain/stress, motion, acceleration etc. By virtue of increased output power and advanced technologies, self-powered systems consisting of some devices with high-power consumption will also be realized in the near future. This review can be useful and helpful for developing effective dielectrics of energy harvesters and sensor which are able to supply sufficient energy with portable electronic devices and to be highly sensitive and multifunctional.

Author Contributions: J.P.L. and J.W.L. contributed equally to this work.

Funding: This research was funded by Samsung Research Funding Center of Samsung Electronics under Project, grant number SRFC-TA1403-51.

Conflicts of Interest: The authors declare no conflict of interest.

References

1. Fan, F.-R.; Tian, Z.Q.; Wang, Z.L. Flexible triboelectric generator! *Nano Energy* **2012**, *1*, 328–334. [CrossRef]
2. Fan, F.-R.; Lin, L.; Zhu, G.; Wu, W.; Zhang, R.; Wang, Z.L. Transparent Triboelectric Nanogenerators and Self-Powered Pressure Sensors Based on Micropatterned Plastic Films. *Nano Lett.* **2012**, *12*, 3109–3114. [CrossRef] [PubMed]
3. Chen, J.; Wang, Z.L. Reviving Vibration Energy Harvesting and Self-Powered Sensing by a Triboelectric Nanogenerator. *Joule* **2017**, *1*, 480–521. [CrossRef]
4. Fan, F.-R.; Tang, W.; Wang, Z.L. Flexible Nanogenerators for Energy Harvesting and Self-Powered Electronics. *Adv. Mater.* **2016**, *28*, 4283–4305. [CrossRef] [PubMed]
5. Fan, F.-R.; Luo, J.; Tang, W.; Li, C.; Zhang, C.; Tian, Z.; Wang, Z.L. Highly transparent and flexible triboelectric nanogenerators: Performance improvements and fundamental mechanisms. *J. Mater. Chem. A* **2014**, *2*, 13219–13225. [CrossRef]
6. Bai, P.; Zhu, G.; Liu, Y.; Chen, J.; Jing, Q.; Yang, W.; Ma, J.; Zhang, G.; Wang, Z.L. Cylindrical Rotating Triboelectric Nanogenerator. *ACS Nano* **2013**, *7*, 6361–6366. [CrossRef] [PubMed]
7. Chen, J.; Huang, Y.; Zhang, N.; Zou, H.; Liu, R.; Tao, C.; Fan, X.; Wang, Z.L. Micro-cable structured textile for simultaneously harvesting solar and mechanical energy. *Nat. Energy* **2016**, *1*, 16138. [CrossRef]
8. Zhang, B.; Zhang, L.; Deng, W.; Jin, L.; Chun, F.; Pan, H.; Gu, B.; Zhang, H.; Lv, Z.; Yang, W.; et al. Self-Powered Acceleration Sensor Based on Liquid Metal Triboelectric Nanogenerator for Vibration Monitoring. *ACS Nano* **2017**, *11*, 7440–7446. [CrossRef] [PubMed]
9. Wang, J.; Zhang, H.; Xie, Y.; Yan, Z.; Yuan, Y.; Huang, L.; Cui, X.; Gao, M.; Su, Y.; Yang, W.; et al. Smart network node based on hybrid nanogenerator for self-powered multifunctional sensing. *Nano Energy* **2017**, *33*, 418–426. [CrossRef]
10. Wang, Z.L.; Chen, J.; Lin, L. Progress in triboelectric nanogenerators as a new energy technology and self-powered sensors. *Energy Environ. Sci.* **2015**, *8*, 2250–2282. [CrossRef]
11. Zhu, G.; Chen, J.; Zhang, T.; Jing, Q.; Wang, Z.L. Radial-arrayed rotary electrification for high performance triboelectric generator. *Nat. Commun.* **2014**, *5*, 3426. [CrossRef] [PubMed]
12. Zhu, G.; Zhou, Y.S.; Bai, P.; Meng, X.S.; Jing, Q.; Chen, J.; Wang, Z.L. A Shape-Adaptive Thin-Film-Based Approach for 50% High-Efficiency Energy Generation through Micro-Grating Sliding Electrification. *Adv. Mater.* **2014**, *26*, 3788–3796. [CrossRef] [PubMed]
13. Zhu, G.; Bai, P.; Chen, J.; Wang, Z.L. Power-generating shoe insole based on triboelectric nanogenerators for self-powered consumer electronics. *Nano Energy* **2013**, *2*, 688–692. [CrossRef]
14. Chen, J.; Zhu, G.; Yang, J.; Jing, Q.; Bai, P.; Yang, W.; Qi, X.; Su, Y.; Wang, Z.L. Personalized Keystroke Dynamics for Self-Powered Human–Machine Interfacing. *ACS Nano* **2015**, *9*, 105–116. [CrossRef] [PubMed]
15. Zheng, L.; Cheng, G.; Chen, J.; Lin, L.; Wang, J.; Liu, Y.; Li, H.; Wang, Z.L. A Hybridized Power Panel to Simultaneously Generate Electricity from Sunlight, Raindrops, and Wind around the Clock. *Adv. Energy Mater.* **2015**, *5*, 1501152. [CrossRef]
16. Lin, Z.; Chen, J.; Li, X.; Zhou, Z.; Meng, K.; Wei, W.; Yang, J.; Wang, Z.L. Triboelectric Nanogenerator Enabled Body Sensor Network for Self-Powered Human Heart-Rate Monitoring. *ACS Nano* **2017**, *11*, 8830–8837. [CrossRef] [PubMed]
17. Wang, Z.L.; Lin, L.; Chen, J.; Niu, S.; Zi, Y. *Triboelectric Nanogenerators*, 1st ed.; Springer International Publishing: Berlin, Germany, 2016.
18. Bae, J.; Lee, J.; Kim, S.; Ha, J.; Lee, B.-S.; Park, Y.; Choong, C.; Kim, J.-B.; Wang, Z.L.; Kim, H.-Y.; et al. Flutter-driven triboelectrification for harvesting wind energy. *Nat. Commun.* **2014**, *5*, 4929–4937. [CrossRef] [PubMed]
19. Zhu, G.; Su, Y.; Bai, P.; Chen, J.; Jing, Q.; Yang, W.; Wang, Z.L. Harvesting Water Wave Energy by Asymmetric Screening of Electrostatic Charges on a Nanostructured Hydrophobic Thin-Film Surface. *ACS Nano* **2014**, *8*, 6031–6037. [CrossRef] [PubMed]
20. Kim, K.N.; Lee, J.P.; Lee, S.-H.; Lee, S.C.; Baik, J.M. Ergonomically designed replaceable and multifunctional triboelectric nanogenerator for a uniform contact. *RSC Adv.* **2016**, *6*, 88526–88530. [CrossRef]
21. Wang, S.; Lin, L.; Wang, Z.L. Nanoscale Triboelectric-Effect-Enabled Energy Conversion for Sustainably Powering Portable Electronics. *Nano Lett.* **2012**, *12*, 6339–6346. [CrossRef] [PubMed]

22. Zhu, G.; Lin, Z.-H.; Jing, Q.; Bai, P.; Pan, C.; Yang, Y.; Zhou, Y.; Wang, Z.L. Toward Large-Scale Energy Harvesting by a Nanoparticle-Enhanced Triboelectric Nanogenerator. *Nano Lett.* **2013**, *13*, 847–853. [CrossRef] [PubMed]
23. Wang, S.; Lin, L.; Xie, Y.; Jing, Q.; Niu, S.; Wang, Z.L. Sliding-Triboelectric Nanogenerators Based on In-Plane Charge-Separation Mechanism. *Nano Lett.* **2013**, *13*, 2226–2233. [CrossRef] [PubMed]
24. Wang, S.; Xie, Y.; Niu, S.; Lin, L.; Liu, C.; Zhou, Y.S.; Wang, Z.L. Maximum Surface Charge Density for Triboelectric Nanogenerators Achieved by Ionized-Air Injection: Methodology and Theoretical Understanding. *Adv. Mater.* **2014**, *26*, 6720–6728. [CrossRef] [PubMed]
25. Li, X.; Liu, M.; Huang, B.; Liu, H.; Hu, W.; Shao, L.-H.; Wang, Z.L. Nanoporous-Gold-Based Hybrid Cantilevered Actuator Dealloyed and Driven by A Modified Rotary Triboelectric Nanogenerator. *Sci. Rep.* **2016**, *6*, 24092. [CrossRef] [PubMed]
26. Chun, J.; Ye, B.U.; Lee, J.W.; Choi, D.; Kang, C.-Y.; Kim, S.-W.; Wang, Z.L.; Baik, J.M. Boosted output performance of triboelectric nanogenerator via electric double layer effect. *Nat. Commun.* **2016**, *7*, 12985. [CrossRef] [PubMed]
27. Niu, S.; Wang, S.; Lin, L.; Liu, Y.; Zhou, Y.S.; Hu, Y.; Wang, Z.L. Theoretical study of contact-mode triboelectric nanogenerators as an effective power source. *Energy Environ. Sci.* **2013**, *6*, 3576–3583. [CrossRef]
28. Seung, W.; Yoon, H.-J.; Kim, T.Y.; Ryu, H.; Kim, J.; Lee, J.-H.; Lee, J.H.; Kim, S.; Park, Y.K.; Park, Y.J.; et al. Nanogenerators: Boosting Power-Generating Performance of Triboelectric Nanogenerators via Artificial Control of Ferroelectric Polarization and Dielectric Properties. *Adv. Energy Mater.* **2017**, *7*, 1600988. [CrossRef]
29. Ye, B.U.; Kim, B.-Y.; Ryu, J.; Lee, J.Y.; Baik, J.M.; Hong, K. Electrospun ion gel nanofibers for flexible triboelectric nanogenerator: Electrochemical effect on output power. *Nanoscale* **2015**, *7*, 16189–16194. [CrossRef] [PubMed]
30. Lee, K.Y.; Yoon, H.-J.; Jiang, T.; Wen, X.; Seung, W.; Kim, S.-W.; Wang, Z.L. Fully Packaged Self-Powered Triboelectric Pressure Sensor Using Hemispheres-Array. *Nano Lett.* **2012**, *12*, 3109–3114. [CrossRef]
31. Dhakar, L.; Gudla, S.; Shan, X.; Wang, Z.; Tay, F.E.H.; Heng, C.H.; Lee, C. Large Scale Triboelectric Nanogenerator and Self-Powered Pressure Sensor Array Using Low Cost Roll-to-Roll UV Embossing. *Sci. Rep.* **2016**, *6*, 22253. [CrossRef] [PubMed]
32. Yang, J.; Chen, J.; Su, Y.; Jing, Q.; Li, Z.; Yi, F.; Wen, X.; Wang, Z.; Wang, Z.L. Eardrum-Inspired Active Sensors for Self-Powered Cardiovascular System Characterization and Throat-Attached Anti-Interference Voice Recognition. *Adv. Mater.* **2015**, *27*, 1316–1326. [CrossRef] [PubMed]
33. Lin, Z.; Yang, J.; Li, X.; Wu, Y.; Wei, W.; Liu, J.; Chen, J.; Yang, J. Large-Scale and Washable Smart Textiles Based on Triboelectric Nanogenerator Arrays for Self-Powered Sleeping Monitoring. *Adv. Funct. Mater.* **2018**, *28*, 1704112. [CrossRef]
34. Yang, W.; Chen, J.; Wen, X.; Jing, Q.; Yang, J.; Su, Y.; Zhu, G.; Wu, W.; Wang, Z.L. Triboelectrification Based Motion Sensor for Human-Machine Interfacing. *ACS Appl. Mater. Interfaces* **2014**, *6*, 7479–7484. [CrossRef] [PubMed]
35. Zhang, B.; Chen, J.; Jin, L.; Deng, W.; Zhang, L.; Zhang, H.; Zhu, M.; Yang, W.; Wang, Z.L. Rotating-Disk-Based Hybridized Electromagnetic–Triboelectric Nanogenerator for Sustainably Powering Wireless Traffic Volume Sensors. *ACS Nano* **2016**, *10*, 6241–6247. [CrossRef] [PubMed]
36. Chen, J.; Zhu, G.; Yang, W.; Jing, Q.; Bai, P.; Yang, Y.; Hou, T.; Wang, Z.L. Harmonic-Resonator-Based Triboelectric Nanogenerator as a Sustainable Power Source and a Self-Powered Active Vibration Sensor. *Adv. Mater.* **2013**, *25*, 6094–6099. [CrossRef] [PubMed]
37. Kim, J.; Chun, J.; Kim, J.W.; Choi, W.J.; Baik, J.M. Self-powered Room Temperature Electronic Nose based on Triboelectrification and Heterogeneous Catalytic Reaction. *Adv. Funct. Mater.* **2015**, *25*, 7049–7055. [CrossRef]
38. Lin, Z.-H.; Cheng, G.; Wu, W.; Pradel, K.C.; Wang, Z.L. Dual-Mode Triboelectric Nanogenerator for Harvesting Water Energy and as a Self-Powered Ethanol Nanosensor. *ACS Nano* **2014**, *8*, 6440–6448. [CrossRef] [PubMed]
39. Zhang, H.; Yang, Y.; Su, Y.; Chen, J.; Hu, C.; Wu, Z.; Liu, Y.; Wong, C.P.; Bando, Y.; Wang, Z.L. Triboelectric nanogenerator as self-powered active sensors for detecting liquid/gaseous water/ethanol. *Nano Energy* **2013**, *2*, 693–701. [CrossRef]
40. Wen, Z.; Chen, J.; Yeh, M.-H.; Guo, H.; Li, Z.; Fan, X.; Zhang, T.; Zhu, L.; Wang, Z.L. Blow-driven triboelectric nanogenerator as an active alcohol breath analyzer. *Nano Energy* **2015**, *16*, 38–46. [CrossRef]

41. Lin, Z.-H.; Zhu, G.; Zhou, Y.S.; Yang, Y.; Bai, P.; Chen, J.; Wang, Z.L. A Self-Powered Triboelectric Nanosensor for Mercury Ion Detection. *Angew. Chem.* **2013**, *52*, 1–6. [CrossRef] [PubMed]
42. Zhang, H.; Yang, Y.; Hou, T.-C.; Su, Y.; Hu, C.; Wang, Z.L. Triboelectric nanogenerator built inside clothes for self-powered glucose biosensors. *Nano Energy* **2013**, *2*, 1019–1024. [CrossRef]
43. Li, Z.; Chen, J.; Yang, J.; Su, Y.; Fan, X.; Wu, Y.; Yu, C.; Wang, Z.L. β-cyclodextrin enhanced triboelectrification for self-powered phenol detection and electrochemical degradation. *Energy Environ Sci.* **2015**, *8*, 887–896. [CrossRef]
44. Li, Z.; Chen, J.; Guo, H.; Fan, X.; Wen, Z.; Yeh, M.-H.; Yu, C.; Cao, X.; Wang, Z.L. Triboelectrification-Enabled Self-Powered Detection and Removal of Heavy Metal Ions in Wastewater. *Adv. Mater.* **2016**, *28*, 2983–2991. [CrossRef] [PubMed]
45. Lee, J.W.; Ye, B.U.; Baik, J.M. Research Update: Recent progress in the development of effective dielectrics for high-output triboelectric nanogenerator. *APL Mater.* **2017**, *5*, 073802. [CrossRef]
46. Bai, P.; Zhu, G.; Zhou, Y.S.; Wang, S.; Ma, J.; Zhang, G.; Wang, Z.L. Dipole-moment-induced effect on contact electrification for triboelectric nanogenerators. *Nano Res.* **2014**, *7*, 990–997. [CrossRef]
47. Soin, N.; Zhao, P.; Prashanthi, K.; Chen, J.; Ding, P.; Zhou, E.; Shah, T.; Ray, S.C.; Tsonos, C.; Thundat, T.; et al. High performance triboelectric nanogenerators based on phase-inversionpiezoelectric membranes of poly(vinylidenefluoride)-zinc stannate (PVDF-ZnSnO$_3$) and polyamide-6 (PA6). *Nano Energy* **2016**, *30*, 470–480. [CrossRef]
48. Lee, J.W.; Cho, H.J.; Chun, J.; Kim, K.N.; Kim, S.; Ahn, C.W.; Kim, I.W.; Kim, J.-Y.; Kim, S.-W.; Yang, C.; et al. Robust nanogenerators based on graft copolymers via control of dielectrics for remarkable output power enhancement. *Sci. Adv.* **2017**, *3*, 1602902. [CrossRef] [PubMed]
49. Chun, J.; Kim, J.W.; Jung, W.; Kang, C.-Y.; Kim, S.-W.; Wang, Z.L.; Baik, J.M. Mesoporous pores impregnated with Au nanoparticles as effective dielectrics for enhancing triboelectric nanogenerator performance in harsh environments. *Energy Environ. Sci.* **2015**, *8*, 3006–3012. [CrossRef]
50. Lee, J.P.; Ye, B.Y.; Kim, K.N.; Lee, J.W.; Choi, W.J.; Baik, J.M. 3D printed noise-cancelling triboelectric nanogenerator. *Nano Energy* **2017**, *38*, 377–384. [CrossRef]
51. Bai, P.; Zhu, G.; Lin, Z.-H.; Jing, Q.; Chen, J.; Zhang, G.; Ma, J.; Wang, Z.L. Integrated Multilayered Triboelectric Nanogenerator for Harvesting Biomechanical Energy from Human Motions. *ACS Nano* **2013**, *7*, 3713–3719. [CrossRef] [PubMed]
52. Su, Y.; Wen, X.; Zhu, G.; Yang, J.; Chen, J.; Bai, P.; Wu, Z.; Jiang, Y.; Wang, Z.L. Hybrid triboelectric nanogenerator for harvesting water wave energy and as a self-powered distress signal emitter. *Nano Energy* **2014**, *9*, 186–195. [CrossRef]
53. Pu, X.; Liu, M.; Li, L.; Zhang, C.; Pang, Y.; Jiang, C.; Shao, L.; Hu, W.; Wang, Z.L. Efficient Charging of Li-Ion Batteries with Pulsed Output Current of Triboelectric Nanogenerators. *Adv. Sci.* **2015**, *26*, 1500255. [CrossRef] [PubMed]
54. Bai, P.; Zhu, G.; Jing, Q.; Yang, J.; Chen, J.; Su, Y.; Ma, J.; Zhang, G.; Wang, Z.L. Membrane-Based Self-Powered Triboelectric Sensors for Pressure Change Detection and Its Uses in Security Surveillance and Healthcare Monitoring. *Adv. Funct. Mater.* **2014**, *24*, 5807–5813. [CrossRef]
55. Xie, Y.; Wang, S.; Niu, S.; Lin, L.; Jing, Q.; Yang, J.; Wu, Z.; Wang, Z.L. Grating-Structured Freestanding Triboelectric-Layer Nanogenerator for Harvesting Mechanical Energy at 85% Total Conversion Efficiency. *Adv. Mater.* **2014**, *26*, 6599–6607. [CrossRef] [PubMed]
56. Lin, L.; Xie, Y.; Niu, S.; Wang, S.; Yang, P.-K.; Wang, Z.L. Robust Triboelectric Nanogenerator Based on Rolling Electrification and Electrostatic Induction at an Instantaneous Energy Conversion Efficiency of ~55%. *ACS Nano* **2015**, *9*, 922–930. [CrossRef] [PubMed]
57. Lin, L.; Wang, S.; Xie, Y.; Jing, Q.; Niu, S.; Hu, Y.; Wang, Z.L. Segmentally Structured Disk Triboelectric Nanogenerator for Harvesting Rotational Mechanical Energy. *Nano Lett.* **2013**, *13*, 2916–2923. [CrossRef] [PubMed]
58. Zhao, Z.; Pu, X.; Du, C.; Li, L.; Jiang, C.; Hu, W.; Wang, Z.L. Freestanding Flag-Type Triboelectric Nanogenerator for Harvesting High-Altitude Wind Energy from Arbitrary Directions. *ACS Nano* **2016**, *10*, 1780–1787. [CrossRef] [PubMed]
59. Kim, J.; Lee, J.H.; Ryu, H.; Lee, J.-H.; Khan, U.; Kim, H.; Kwak, S.S.; Kim, S.-W. High-Performance Piezoelectric, Pyroelectric, and Triboelectric Nanogenerators Based on P(VDF-TrFE) with Controlled Crystallinity and Dipole Alignment. *Adv. Funct. Mater.* **2017**, *27*, 1700702. [CrossRef]

60. Huang, T.; Lu, M.; Yu, H.; Zhang, Q.; Wang, H.; Zhu, M. Enhanced Power Output of a Triboelectric Nanogenerator Composed of Electrospun Nanofiber Mats Doped with Graphene Oxide. *Sci. Rep.* **2015**, *5*, 13942. [CrossRef] [PubMed]
61. Zhu, Y.; Yang, B.; Liu, J.; Wang, X.; Wang, L.; Chen, X.; Yang, C. A flexible and biocompatible triboelectric nanogenerator with tunable internal resistance for powering wearable devices. *Sci. Rep.* **2016**, *6*, 22233. [CrossRef] [PubMed]
62. Lin, Z.-H.; Cheng, G.; Yang, Y.; Zhou, Y.S.; Lee, S.; Wang, Z.L. Triboelectric Nanogenerator as an Active UV Photodetector. *Adv. Funct. Mater.* **2014**, *24*, 2810–2816. [CrossRef]
63. Kwon, Y.H.; Shin, S.-H.; Kim, Y.-H.; Kim, J.-Y.; Lee, M.H.; Nah, J. Triboelectric contact surface charge modulation and piezoelectric charge inducement using polarized composite thin film for performance enhancement of triboelectric generators. *Nano Energy* **2016**, *25*, 225–231. [CrossRef]
64. Chen, J.; Guo, H.; He, X.; Liu, G.; Xi, Y.; Shi, H.; Hu, C. Enhancing Performance of Triboelectric Nanogenerator by Filling High Dielectric Nanoparticles into Sponge PDMS Film. *ACS Appl. Mater. Interfaces* **2016**, *8*, 736–744. [CrossRef] [PubMed]
65. Zheng, Y.; Cheng, L.; Yuan, M.; Wang, Z.; Qin, Y.; Wang, T. An electrospun nanowire-based triboelectric nanogenerator and its application in a fully self-powered UV detector. *Nanoscale* **2014**, *6*, 7842–7846. [CrossRef] [PubMed]
66. Wang, X.; Yang, B.; Liu, J.; Zhu, Y.; Yang, C.; He, Q. A flexible triboelectric-piezoelectric hybrid nanogenerator based on P(VDF-TrFE) nanofibers and PDMS/MWCNT for wearable devices. *Sci. Rep.* **2016**, *6*, 36409. [CrossRef] [PubMed]
67. Park, J.; Kim, M.; Lee, Y.; Lee, H.S.; Ko, H. Fingertip skin–inspired microstructured ferroelectric skins discriminate static/dynamic pressure and temperature stimuli. *Sci. Adv.* **2015**, *1*, 1500661. [CrossRef] [PubMed]
68. Lee, J.S.; Shin, K.-Y.; Cheong, O.J.; Kim, J.H.; Jang, J. Highly sensitive and multifunctional tactile sensor using free-standing ZnO/PVDF thin film with graphene electrodes for pressure and temperature monitoring. *Sci. Rep.* **2015**, *5*, 7887. [CrossRef] [PubMed]
69. Lin, M.-F.; Xiong, J.; Wang, J.; Parida, K.; Lee, P.S. Core-shell nanofiber mats for tactile pressure sensor and nanogenerator applications. *Nano Energy* **2018**, *44*, 248–255. [CrossRef]

© 2018 by the authors. Licensee MDPI, Basel, Switzerland. This article is an open access article distributed under the terms and conditions of the Creative Commons Attribution (CC BY) license (http://creativecommons.org/licenses/by/4.0/).

Development of the Triboelectric Nanogenerator Using a Metal-to-Metal Imprinting Process for Improved Electrical Output

Moonwoo La [1], Jun Hyuk Choi [2], Jeong-Young Choi [1], Taek Yong Hwang [1], Jeongjin Kang [1] and Dongwhi Choi [2,*]

1. Molds & Dies Technology R&D Group, Korea Institute of Industrial Technology (KITECH), Incheon 21999, Korea; mla@kitech.re.kr (M.L.); wjddud21@kitech.re.kr (J.-Y.C.); taekyong@kitech.re.kr (T.Y.H.); doublej@kitech.re.kr (J.K.)
2. Department of Mechanical Engineering, Kyung Hee University, Yongin 17104, Korea; cjng123@khu.ac.kr
* Correspondence: dongwhi.choi@khu.ac.kr; Tel.: +82-31-201-3694

Received: 1 October 2018; Accepted: 23 October 2018; Published: 27 October 2018

Abstract: Triboelectric nanogenerators (TENG), which utilize contact electrification of two different material surfaces accompanied by electrical induction has been proposed and is considered as a promising energy harvester. Researchers have attempted to form desired structures on TENG surfaces and successfully demonstrated the advantageous effect of surface topography on its electrical output performance. In this study, we first propose the structured Al (SA)-assisted TENG (SA-TENG), where one of the contact layers of the TENG is composed of a structured metal surface formed by a metal-to-metal (M2M) imprinting process. The fabricated SA-TENG generates more than 200 V of open-circuit voltage and 60 µA of short-circuit current through a simple finger tapping motion. Given that the utilization of the M2M imprinting process allows for the rapid, versatile and easily accessible structuring of various metal surfaces, which can be directly used as a contact layer of the TENG to substantially enhance its electrical output performance, the present study may considerably broaden the applicability of the TENG in terms of its fabrication standpoint.

Keywords: triboelectric nanogenerator; nanoimprinting; nanostructures; microstructures; femtosecond laser

1. Introduction

The development of portable electronic devices has received considerable attention due to their unique advantages such as convenience and excellent functionality, which has enhanced the lives of people in various aspects. These devices require power without using wires from an external power source. As a result, energy-harvesting technologies, which convert the available sustainable energies into electricity has also attracted attention worldwide [1]. The eminent concept of the triboelectric nanogenerator (TENG), which utilizes contact electrification of two different material surfaces accompanied by electrical induction, has been proposed [2]. Since its first proposal in 2012, TENG has been actively studied, hence, it is considered to be a promising energy harvesting technology to operate portable electronic devices without spatio-temporal limitations on the basis of its advantages, such as material selection diversity, high efficiency and high shape adaptability [3–8]. The most actively conducted research topic in the field of the TENG is the enhancement of its electrical output performance, which is similar to those of other energy-harvesting technologies [9]. Given that the fundamental operating mechanism of the TENG is based on the contact and separation between two surfaces, the simplest and most widely used strategy to enhance the electrical output performance is the introduction of micro- and nanoscale structures onto the surfaces where friction occurs [3,5,10–12].

The formation of the surface structures significantly increases local contact pressure resulting in further increases in the contact area between two contact surfaces, thereby generating a high amount of electrical charge on the contact surface [13]. Given that the most widely utilized material is based on polymer, many researchers have attempted to form desired structures onto polymer surfaces through various subtractive and additive fabrication methodologies. These researchers have successfully demonstrated the benefits of surface topography on the electrical output performance of the TENG. Although the electrical output performance of the TENG has drastically increased due to these efforts, the introduction of structures on the polymer surface accelerates the mechanical wear of the surface of the friction due to the increased local contact pressure. Considering that the operation of a TENG is based on the friction between two surfaces, one of the major issues to be resolved is the problem of mechanical wear [5,14,15]. However, the introduction of structures onto the polymer has relatively weak mechanical characteristics compared with other engineering materials, which is definitely unfavorable in terms of mechanical wear. A potential strategy to reduce mechanical wear in the TENG can be the formation of the structures on metal surfaces instead of polymer surfaces. Metal is a widely utilized engineering material with better mechanical characteristics, thereby resulting in higher abrasion resistance than those of the polymer. Metal is also widely utilized in the TENG causing friction with the polymer surface. The common approach to the fabrication of small-scale structures on metal is first to fabricate a prepatterned polymer and use it as a sacrificial layer. Then, subsequent metal deposition on its surface followed by lift-off or an etching process, which are considered to be complex and labor intensive multistep processes due to their requirement for difficult processing conditions [16,17]. Consequently, most TENGs, which utilize a polymer-metal friction to generate electrical output performance, only have a structure on the polymer surface because it takes considerable effort to form structures on the metal surface than on the polymer surface [3]. Hence, the proposal of facile and highly accessible strategies to form structures onto the metal surface will open another horizon to enhance the electrical output performance of the TENG.

In this study, we propose a structured Al (SA)-assisted TENG (SA-TENG), where one of the contact surfaces of the TENG is composed of the structured metal surface formed by a metal-to-metal (M2M) imprinting process. The imprinting process is a low-cost and rapid process that involves transcribing structures onto the substrate of interest by using a stamp. The present study utilizes a precise femtosecond (fs) laser to fabricate steel stamps, which have optically induced micro- and nanoscale patterns and these patterns are transcribed on Al substrates by applying heat and pressure. The M2M imprinting process is optimized by applying a systematic approach. As a result, conical microstructures and line nanostructures are successfully transcribed onto the Al substrates. The structures on the Al substrates considerably enhance the electrical output performance of the present SA-TENG. Given that the utilization of the M2M imprinting process allows the rapid, versatile and easily accessible structuring of the metal substrate, which can be directly used as a contact layer on the TENG to significantly enhance its electrical output performance, the present study may significantly broaden the applicability of the TENG in terms of its fabrication standpoint.

2. Materials and Methods

2.1. Fabrication of Stamps via fs Laser Processing of Steels

The laser used for stamp fabrication was a Ti:sapphire fs laser system on the basis of chirped pulse amplification. This laser system generated 120 fs pulses with a maximum energy of 5 mJ/pulse and operated at a central wavelength of 800 nm with a repetition rate of 1 kHz. Prior to fs laser processing, we polished steel plates mechanically by using 80 nm-grade colloidal silica until the average surface roughness (Ra) reached 4.1 nm [18]. This polishing process can minimize non-uniform laser energy absorption at the surface due to scratches and pre-existing surface structures [19].

By irradiating the steel plates with fs laser pulses, two types of stamps each with laser-induced conical microstructures and periodic line nanostructures were fabricated on steel (STAVAX, ASSAB,

Incheon, South Korea). To produce conical microstructures, circularly polarized fs laser pulses were used so that each conical microstructure tended to form circularly at the steel surface [20] and the pulses were weakly focused onto the surface with a $1/e^2$ intensity spot radius of 270 µm at a fluence of 0.37 J/cm². The speed of raster scanning was 2 mm/s and the distance between scanning lines was 30 µm. On the other hand, to create periodic line nanostructures, linearly polarized fs laser pulses were used, since the orientation (wave vector) of periodic line nanostructures is parallel to laser polarization [21]. Compared to the case of conical microstructure formation, fs laser pulses were rather tightly focused with a $1/e^2$ intensity spot radius of 90 µm. Additionally, a lower laser fluence, faster raster scanning speed and larger scanning line distance of 0.23 J/cm², 12 mm/s and 40 µm were used for the fabrication of periodic line nanostructures. All experiments in this sub-section were performed in ambient air at normal incidence.

2.2. The SA Fabrication by M2M Imprinting Process

SA fabrication is based on the M2M imprinting process. An as-fabricated steel stamp was utilized and the conventional Al plate was first electropolished and then utilized as a substrate in the M2M imprinting process. The Al substrate and the steel stamp with structures to be transcribed were vertically stacked, and appropriate pressure and heat were applied using a conventional hot-pressing machine (QM900, QMESYS, Gunpo, South Korea).

2.3. Measurement of the Electrical Output Performance of the TENG

A vertical cyclic force was applied on the contact surfaces of the TENG by using the conventionally available vibration testing machine (KD-9363ED-41E, Kingdesign, New Taipei City, Taiwan). The open circuit voltage (V_{OC}), which assumed the infinite load resistance connected to the TENG, was measured by directly connecting the oscilloscope (DS1074z, Rigol, Beaverton, OR, USA) equipped with the high voltage probe (DP-50, Pintek, New Taipei City, Taiwan) to the TENG. The internal resistances of the oscilloscope and the high voltage probe were 10 MΩ and 54 MΩ, respectively. The short circuit current (I_{SC}) was measured with a low-noise current preamplifier (SR570, Stanford Research Systems, Sunnyvale, CA, USA) that was connected with an oscilloscope.

3. Results

3.1. Fabrication of the SA

Figure 1 shows the SA fabrication procedure, which is composed of two steps, as follows: (1) the fabrication of the high-strength steel stamp with structures using a fs laser and (2) as-fabricated stamp utilization to transcribe structures on the Al substrate by using the M2M imprinting process. Since its development, the fs laser has become one of the most promising tools for direct surface patterning on various engineering materials, including metals. Many reports have shown that the interaction between the laser spot and metal surfaces induces natural consequences with the production of nano- to microstructures on the metal surface. These structures also modify the surface properties [17–21]. According to the literature, steel substrates are utilized as engineering materials, nano- and microstructures are easily formed and then as-prepared steel plates are directly utilized as metal stamps in the M2M imprinting process [17]. The M2M imprinting process directly transcribes the structures onto the steel stamp and then to the other metal substrates by applying heat and pressure. In this study, considering that the stamp was composed of steel, which has relatively high strength, most of the other conventionally available metals, such as Al, Cu, Ag and Au, can be utilized as a target substrate. Here, considering that Al is often utilized as one of the contact layers of the TENG due to its low electron affinity, we exclusively utilizes the Al plate as our substrate, where the transcription of the structures occurred. After placing the steel stamp and the Al plate up and down, heat and pressure were applied using a hot-pressing machine. Then, the patterns on the stamp was transcribed onto the Al plate.

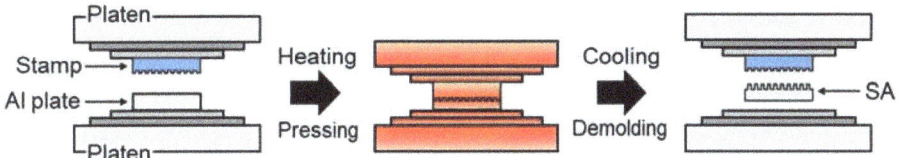

Figure 1. A schematic illustration of the metal-to-metal (M2M) imprinting process. An as-prepared steel stamp and the Al plate is vertically stacked between two platens. Heating and pressing induces transcription of the structures onto the Al plate and following this, cooling and demolding enables detachment of the structured Al (SA) from the stamp.

Figure 2a-(i) shows the images of the fabricated steel stamp with conical microstructures on its surface. The processing conditions to form these structures were based on the literature on the formation of laser-induced microstructures on metal surfaces. The details are provided in the Materials and Methods section. The scanning electron microscopic (SEM, FEI Quanta 200F, ThermoFisher Scientific, MA, USA) image as shown in Figure 2a-(ii) demonstrates the successful formation of conical microstructures on the steel stamp by using the fs laser. Prior to the M2M imprinting process, the features on the stamp surface were investigated using atomic force microscopy (AFM, Park XE-100, Park Systems, Suwon, South Korea) for quantitative analyses, as shown in Figure 2a-(iii). Three representative parameters, that is, width (W), period (P) and height (H) of the structures describing the features on the surface were measured and then utilized to investigate the effect of the processing conditions of the M2M imprinting on the transcription qualities (TQs). This is shown as follows:

$$TQ_W = \frac{W_{substrate}}{W_{stamp}}, \quad TQ_P = \frac{P_{substrate}}{P_{stamp}}, \quad TQ_H = \frac{H_{substrate}}{H_{stamp}} \quad (1)$$

In general, the imprinting pressure (P_i), imprinting temperature (T_i), and holding time (t_i) are utilized as the control parameters of the M2M imprinting process. TQs are investigated by varying these parameters, hence, one of the conditions (P_i = 15 MPa, T_i = 200 °C, t_i = 1 min), where the features of conical microstructures are transcribed well onto the Al substrate, is determined (see Table S1–S3 in Supplementary Materials). Figure 2b-(i–iii) show the structured Al substrate, SEM and AFM images, respectively.

To show the ability of the M2M imprinting process for nanoscale structure transcription, we fabricated the steel stamp possessing periodic line nanostructures. The fabrication procedures were based on a previous study and is also described in the Section 2. As shown in Figure 2c-(i), the size of the features on the stamp is comparable with the range of the period of the visible lights and the stamp exhibits a structural color. The SEM images shown in Figure 2c-(ii) show that the laser-induced periodic line nanostructures were formed well on the steel stamp surface. Experiments and analyses in terms of TQs were conducted in a similar manner to the experiment above. Thus, the condition (i.e., P_i = 25 MPa, T_i = 200 °C, t_i = 5 min) where the features of line nanostructures were transcribed on the Al substrate well were determined (see Table S4 in Supplementary Materials). As shown in Figure 2d-(i), the transcribed nanoscale structures on the Al substrate showed structural color. This result directly showed that the present M2M process could be applied to form a variety of patterns in nanometer and micrometer scales on the Al substrate.

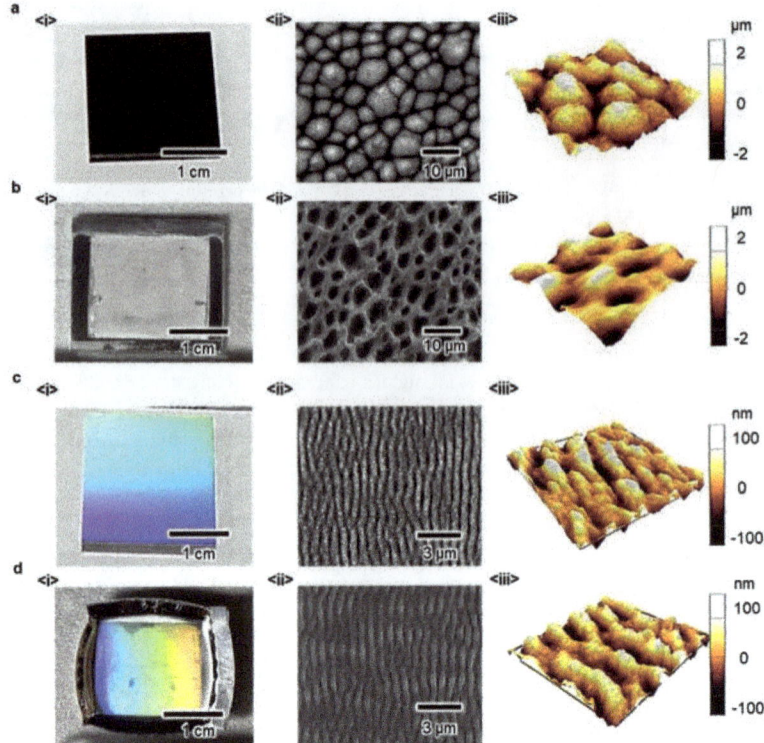

Figure 2. Images and investigation results of the as-prepared steel stamp and the imprinted SA. (**a**) (i) real, (ii) scanning electron microscopic (SEM) and (iii) atomic force microscopy (AFM) images of the steel stamp possessing conical microstructures, (**b**) (i) real, (ii) SEM and (iii) AFM images of the SA with microstructures, (**c**) (i) real, (ii) SEM and (iii) AFM images of the steel stamp possessing line nanostructures, (**d**) (i) real, (ii) SEM and (iii) AFM images of the SA with line nanostructures.

3.2. Utilization of the SA and Resultant Enhanced Electrical Output Performance of the TENG

As-prepared SA possessing micro- and nanoscale structures are utilized as one of the contact layers of the TENG. The electricity generation mechanism of the TENG is shown in Figure 3a. The counter contact layer of the SA is composed of conventional fluorinated ethylene propylene (FEP) film that is attached to the Al film. Given the difference in the electron affinity between FEP and Al, contact spontaneously generates positive and negative electrical charges on Al and FEP, respectively. The separation of the two surfaces induces generation of the net electrical charges on both surfaces and corresponding charge generation on the FEP film surface induces the additional positive charges on the Al film beneath the FEP film. During this process, the electrical current, which is generated from the movement of the electrical charges, flows. Another contact of Al with FEP satisfies the electro-neutrality between the electrical charges on Al and FEP. Hence, electrical current flows in the opposite direction. Consequently, repetitive contact and separation between Al and FEP result in alternating current, as shown in Figure 3a.

To investigate the effect of the features on the electrical output performance of the TENG, we utilized flat Al (control), SA with conical microstructures (SA_{micro}) and SA with line nanostructures (SA_{nano}) for comparison. Under the same experimental conditions, the $V_{OC}s$ and the $I_{SC}s$ of each case were measured. Vertical cyclic force with a magnitude of 8 N and a frequency of 10 Hz was applied on the contact surfaces by using the conventionally available vibration testing machine.

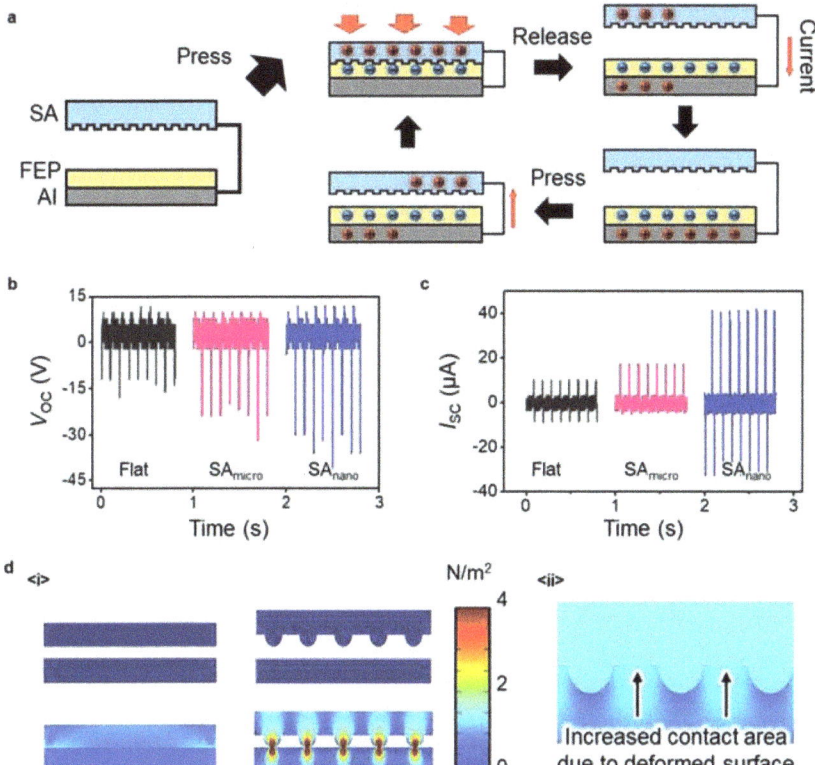

Figure 3. Schematics of the operating mechanism of the SA-TENG and its electrical output performance. (**a**) Operating mechanism of the SA-TENG. Successive contact and separation between FEP and Al surfaces generates electrical charges and corresponding electric current flows, (**b**) comparison of the V_{OC}s and (**c**) I_{SC}s generated from the TENGs with flat Al, SA_{micro} and SA_{nano}, (**d**) (i) numerical analysis showing the increased local contact pressure on the structured surfaces, (ii) the deformation of the surface due to the increased local contact pressure which increases the total contact area, resulting in an increase in the amount of electrical charges generated on the surface.

As shown in Figure 3b,c, both V_{OC}s and I_{SC}s generated from SA_{micro} and SA_{nano} are larger than those from the control. The close-up views of the V_{OC} and I_{SC} show the detailed profiles of the peaks (see Figure S1 in Supplementary Materials). The result in Figure 3b,c is consistent with the previously reported findings with the various TENGs possessing surface structures. The structures on the surface are known to increase the local pressure during contact situations resulting in further increases in the contact area between two contact surfaces, thereby increasing the amount of electrical charges generated on the surface [13]. The amount of electrical charges on the surface is positively correlated with the electrical output performance of the TENG. This claim can be verified using the numerical simulation shown in Figure 3d. To investigate the effect of the surface structures on the developed localized pressure during contact, we conducted a numerical simulation by using the commercial software COMSOL Multiphysics™ (COMSOL Inc., Burlington, MA, USA). The simplified models of the flat and structured Al surfaces are shown in Figure 3d-(i). When we apply a force, the highly increased local contact pressure can be observed on the structured surface compared to that on the flat surface. This increased local contact pressure further induces deformation of the contact surface, resulting in an increase in the contact area between two surfaces as shown in Figure 3d-(ii). Details on the numerical analysis of the surface deformation are in the Figure S2 in Supplementary Materials. The result directly supports the claim that the SA is preferable in enhancing the electrical output performance of the TENG due to the presence of the structures on its surface.

Considering that metal is a widely utilized engineering material with better mechanical characteristics with higher endurance than those of the polymer, one of the benefits of using the structured metal surface as one of the layers of the TENG would be the stable electrical output performance. To support such a claim, the degradation behavior of the electrical output performance as well as the endurance of the structures on the present SA under an extremely large number of the contact separation working cycles were investigated. It is noteworthy to mention that there is no noticeable degradation of the electrical output performance and significant change of the surface structures after ~10^6 working cycles (see Figure S3 in Supplementary Materials). The experimental results directly showed that the formation of the structures on metal in the present study could be a potent strategy not only to enhance the electrical output performance of the TENG, but also to increase the endurance of the device.

3.3. Fabrication of the SA-TENG

On the basis of the results in the previous section, we fabricated the SA-TENG, which included a counter layer composed of FEP and Al films in the previous experiment via SA_{nano} and as shown in Figure 4a. The FEP film attached onto the Al film and as-prepared SA_{nano} were utilized as two separated components of the SA-TENG, which contacted and separated with each other to generate electricity. The two components were attached and supported by the polydimethylsiloxane-attached poly(methyl methacrylate) plate, which was elaborately carved by the laser cutter. Four compressive springs located at the vertices of the rectangular plate maintained the position of the components with a separated status through spring resilience. When we pressed the upper side of the SA-TENG, the exerted compressive force allowed for the contact of SA and FEP. Thus, the electricity started to generate based on the electricity generation principle mentioned above. With the present SA-TENG, the single finger pressing motion can generate 200 V of V_{OC} and 60 µA of I_{SC}. This amount of electricity is sufficient to directly light tens of the conventional light emitting diodes (LEDs) simultaneously (Figure 4b). The letters M, A, P, L, A and b, which are composed of parallel connected LEDs, can easily be lit. The harvested energy was stored in the conventional capacitor, which directly showed that the present SA-TENG could generate electricity from human body motion for the operation of the portable electronic devices. As shown in Figure 4c, the voltage of the capacitor (10 µF) reached 1.5 V within 8 s under vertical cyclic force with a magnitude of 8 N and a frequency of 10 Hz. Hence, the present strategy for the enhanced electrical output performance of TENG by using the M2M imprinting process is expected to widen the applicability of the TENG as a potent and practical energy harvester to power portable and wearable electronic devices.

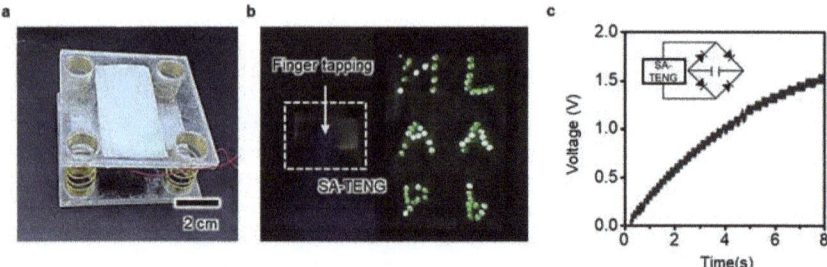

Figure 4. A fabricated SA-TENG and its demonstration of operation. The electrical output generated from the SA-TENG by a simple finger tapping motion enables simultaneous lighting up of the tens of LEDs. Applying the vertical cyclic force on the device charges the conventional capacitor.

4. Discussion

In this study, the steel stamps, which have nanometer or micrometer scale surface structures are successfully fabricated by using an fs laser. The fabricated steel stamps were then directly utilized in the M2M imprinting process to pattern the Al substrate, which can be directly utilized as one of the contact layers of the SA-TENG. The transcribed structures on the Al substrate were shown to play a role in enhancing the electrical output performance of the TENG through corresponding increased local pressure. By using the fabricated SA-TENG, the 200 V of V_{OC} and 60 μA of I_{SC} were generated through single finger press motion. Given that the utilization of the M2M imprinting process enables rapid, versatile and easily accessible structuring of the metal substrate, which can be directly used as a contact layer on the TENG to significantly enhance its electrical output performance, the present study might significantly contribute to broadening the applicability of the TENG from a fabrication standpoint.

Supplementary Materials: The following materials are available online at http://www.mdpi.com/2072-666X/9/11/551/s1, Table S1: M2M processing condition and design of experiment, Table S2: Nine representative cases of experiments in the M2M imprinting process, Table S3: *TQ*s in terms of conical microstructures, Table S4: *TQ*s in terms of line nanostructures. Figure S1: Close-up views of the V_{OC} and I_{SC} showing the detailed profiles of peaks. Figure S2: Result of numerical analysis to investigate the deformation of the contact surface during a contact situation. Figure S3: Experimental result showing strong mechanical durability of the present SA-TENG.

Author Contributions: Data curation, J.K.; investigation, M.L. and J.H.C.; methodology, J.-Y.C. and T.Y.H.; supervision, D.C.; writing—original draft, M.L., J.H.C. and D.C.

Funding: This work was supported by EO180024 as part of KITECH's R&D programs. This research was funded by the grant from the Kyung Hee University in 2018. (KHU-20181163) and from the National Research Foundation of Korea (NRF) grant funded by the Korea government (MSIP). (No. 2016R1A6A3A11931103).

Conflicts of Interest: The authors declare no conflicts of interest.

References

1. Beeby, S.P.; Tudor, M.J.; White, N. Energy harvesting vibration sources for microsystems applications. *Meas. Sci. Technol.* **2006**, *17*, R175. [CrossRef]
2. Fan, F.-R.; Tian, Z.-Q.; Wang, Z.L. Flexible triboelectric generator. *Nano Energy* **2012**, *1*, 328–334. [CrossRef]
3. Zhu, G.; Lin, Z.-H.; Jing, Q.; Bai, P.; Pan, C.; Yang, Y.; Zhou, Y.; Wang, Z.L. Toward large-scale energy harvesting by a nanoparticle-enhanced triboelectric nanogenerator. *Nano Lett.* **2013**, *13*, 847–853. [CrossRef] [PubMed]
4. Choi, D.; Lee, S.; Park, S.M.; Cho, H.; Hwang, W.; Kim, D.S. Energy harvesting model of moving water inside a tubular system and its application of a stick-type compact triboelectric nanogenerator. *Nano Res.* **2015**, *8*, 2481–2491. [CrossRef]
5. Choi, D.; Yoo, D.; Kim, D.S. One-step fabrication of transparent and flexible nanotopographical-triboelectric nanogenerators via thermal nanoimprinting of thermoplastic fluoropolymers. *Adv. Mater.* **2015**, *27*, 7386–7394. [CrossRef] [PubMed]
6. Jao, Y.-T.; Yang, P.-K.; Chiu, C.-M.; Lin, Y.-J.; Chen, S.-W.; Choi, D.; Lin, Z.-H. A textile-based triboelectric nanogenerator with humidity-resistant output characteristic and its applications in self-powered healthcare sensors. *Nano Energy* **2018**, *50*, 513–520. [CrossRef]
7. Choi, D.; Yoo, D.; Cha, K.J.; La, M.; Kim, D.S. Spontaneous occurrence of liquid-solid contact electrification in nature: Toward a robust triboelectric nanogenerator inspired by the natural lotus leaf. *Nano Energy* **2017**, *36*, 250–259. [CrossRef]
8. Bhatia, D.; Lee, J.; Hwang, H.J.; Baik, J.M.; Kim, S.; Choi, D. Design of mechanical frequency regulator for predictable uniform power from triboelectric nanogenerators. *Adv. Energy Mater.* **2018**, *8*, 1702667. [CrossRef]
9. Askari, H.; Khajepour, A.; Khamesee, M.B.; Saadatnia, Z.; Wang, Z.L. Piezoelectric and triboelectric nanogenerators: Trends and impacts. *Nano Today* **2018**, *22*, 10–13. [CrossRef]
10. Zhang, X.-S.; Han, M.-D.; Wang, R.-X.; Zhu, F.-Y.; Li, Z.-H.; Wang, W.; Zhang, H.-X. Frequency-multiplication high-output triboelectric nanogenerator for sustainably powering biomedical microsystems. *Nano Lett.* **2013**, *13*, 1168–1172. [CrossRef] [PubMed]

11. Jeong, C.K.; Baek, K.M.; Niu, S.; Nam, T.W.; Hur, Y.H.; Park, D.Y.; Hwang, G.-T.; Byun, M.; Wang, Z.L.; Jung, Y.S. Topographically-designed triboelectric nanogenerator via block copolymer self-assembly. *Nano Lett.* **2014**, *14*, 7031–7038. [CrossRef] [PubMed]
12. Seung, W.; Gupta, M.K.; Lee, K.Y.; Shin, K.-S.; Lee, J.-H.; Kim, T.Y.; Kim, S.; Lin, J.; Kim, J.H.; Kim, S.-W. Nanopatterned textile-based wearable triboelectric nanogenerator. *ACS Nano* **2015**, *9*, 3501–3509. [CrossRef] [PubMed]
13. Seol, M.-L.; Lee, S.-H.; Han, J.-W.; Kim, D.; Cho, G.-H.; Choi, Y.-K. Impact of contact pressure on output voltage of triboelectric nanogenerator based on deformation of interfacial structures. *Nano Energy* **2015**, *17*, 63–71. [CrossRef]
14. Yang, Y.; Zhu, G.; Zhang, H.; Chen, J.; Zhong, X.; Lin, Z.-H.; Su, Y.; Bai, P.; Wen, X.; Wang, Z.L. Triboelectric nanogenerator for harvesting wind energy and as self-powered wind vector sensor system. *ACS Nano* **2013**, *7*, 9461–9468. [CrossRef] [PubMed]
15. Wang, Z.L. Triboelectric nanogenerators as new energy technology and self-powered sensors–principles, problems and perspectives. *Faraday Discuss.* **2015**, *176*, 447–458. [CrossRef] [PubMed]
16. Chou, S.Y.; Krauss, P.R.; Renstrom, P.J. Nanoimprint lithography. *J. Vac. Sci. Technol. B* **1996**, *14*, 4129–4133. [CrossRef]
17. Chen, H.; Chuang, S.; Cheng, H.; Lin, C.; Chu, T. Directly patterning metal films by nanoimprint lithography with low-temperature and low-pressure. *Microelectron. Eng.* **2006**, *83*, 893–896. [CrossRef]
18. Kim, H.J.; Jeong, H.D.; Lee, E.; Shin, Y. Pad surface characterization and its effect on the tribological state in chemical mechanical polishing. *Key Eng. Mater.* **2004**, *257*, 383–388. [CrossRef]
19. Vorobyev, A.Y.; Guo, C. Femtosecond laser nanostructuring of metals. *Opt. Express* **2006**, *14*, 2164–2169. [CrossRef] [PubMed]
20. Hwang, T.Y.; Guo, C. Polarization and angular effects of femtosecond laser-induced conical microstructures on Ni. *J. Appl. Phy.* **2012**, *111*, 083518. [CrossRef]
21. Hwang, T.Y.; Guo, C. Angular effects of nanostructure-covered femtosecond laser induced periodic surface structures on metals. *J. Appl. Phys.* **2010**, *108*, 073523. [CrossRef]

© 2018 by the authors. Licensee MDPI, Basel, Switzerland. This article is an open access article distributed under the terms and conditions of the Creative Commons Attribution (CC BY) license (http://creativecommons.org/licenses/by/4.0/).

Article

Superhydrophobic Water-Solid Contact Triboelectric Generator by Simple Spray-On Fabrication Method

Jihoon Chung [†], Deokjae Heo [†], Banseok Kim and Sangmin Lee [*]

School of Mechanical Engineering, Chung-Ang University, 84, Heukseok-ro, Dongjak-gu, Seoul 06974, Korea; jihoon@cau.ac.kr (J.C.); ejrwo472@naver.com (D.H.); show8910@naver.com (B.K.)
* Correspondence: slee98@cau.ac.kr; Tel.: +82-2-820-5071
† These authors contributed equally to this work.

Received: 26 October 2018; Accepted: 10 November 2018; Published: 13 November 2018

Abstract: Energy harvesting is a method of converting energy from ambient environment into useful electrical energy. Due to the increasing number of sensors and personal electronics, energy harvesting technologies from various sources are gaining attention. Among energy-harvesting technologies, triboelectric nanogenerator (TENG) was introduced as a device that can effectively generate electricity from mechanical motions by contact-electrification. Particularly, liquid-solid contact TENGs, which use the liquid itself as a triboelectric material, can overcome the inevitable friction wear between two solid materials. Using a commercial aerosol hydrophobic spray, liquid-solid contact TENGs, with a superhydrophobic surface (contact angle over 160°) can be easily fabricated with only a few coating processes. To optimize the fabrication process, the open-circuit voltage of sprayed superhydrophobic surfaces was measured depending on the number of coating processes. To demonstrate the simple fabrication and applicability of this technique on random 3D surfaces, a liquid-solid contact TENG was fabricated on the brim of a cap (its complicated surface structure is due to the knitted strings). This simple sprayed-on superhydrophobic surface can be a possible solution for liquid-solid contact TENGs to be mass produced and commercialized in the future.

Keywords: energy harvesting; triboelectric nanogenerator; superhydrophobic surface; spray method; mechanical energy

1. Introduction

Energy harvesting is a method of converting energy from ambient environment into useful electrical energy. Harvesting energy from the ambient environment is gaining more and more interest due to the increasing number of sensors inside personal electronic devices, which consume extra power and drain batteries much faster. In this respect, there is an increasing number of studies on the use of solar [1], thermal [2] and RF [3] energy to power various sensors and electrical components. Among these energy sources, mechanical energy is one of the desirable sources that is not affected by external environment such as weather, location and so forth. To produce electricity from a mechanical input, piezoelectric [4,5], capacitive [6] and electromagnetic transduction [7] generators have been developed. Recently, the triboelectric nanogenerator (TENG) was introduced as a solution that can effectively generate electricity from mechanical motions by contact electrification [8–12]. In particular, liquid-solid contact TENGs, which use the liquid itself as a triboelectric material, are in the spotlight because they can overcome the inevitable friction wear between two solid materials in conventional TENGs [13–15]. For liquid-solid contact TENGs to produce a sustainable energy output, the solid surface must be superhydrophobic to repel the liquid after it falls. However, producing a superhydrophobic coating on metal or polymer surfaces requires complex fabrication procedures such as vapor deposition [16], plasma treatment [17], or self-assembled monolayer coating [18] to create micro-/nanostructures and to lower the surface energy. In addition, these methods have limited applications for complex 3D surfaces. Therefore, for liquid-solid contact TENGs to be commercialized, the fabrication procedures need to be simple and appropriate for mass production.

In this work, we introduce a liquid-solid contact TENG with a superhydrophobic surface fabricated through a spray-on method; this method can also simply produce a superhydrophobic coating on a complex 3D surface. With a commercial aerosol spray, a superhydrophobic surface that has contact angle of over 160° is created just after few coatings. With a simple superhydrophobic surface, the sprayed-on TENG can generate about 30 V per water drop. To optimize the fabrication process, the open-circuit voltage of the sprayed superhydrophobic surface depending on the number of coating process was measured. The sprayed superhydrophobic coating was shown to be maintained even after being subjected to 20 h of water drops (every 0.5 s). The sprayed superhydrophobic surface was able to produce an average peak open-circuit voltage (V_{OC}) of 13.4 V and a closed-circuit current (I_{CC}) of 2.1 µA under continuous water spraying from a commercial shower head. To demonstrate the simple fabrication and applicability on a random 3D surface, a liquid-solid contact TENG was fabricated on the brim of a cap, which has a complicated surface structure due to the knitted strings. The sprayed-on TENG cap produced sufficient electrical output to light up an LED. The superhydrophobic surface created by the aerosol spray method presented in this study can be easily applied to 3D surfaces using a simple fabrication process. Thus, this technique can be a possible solution for liquid-solid contact TENGs to be mass produced in the future.

2. Materials and Methods

2.1. The Fabrication Process of Sprayed-On Superhydrophobic Surface

First, a 10 cm × 10 cm bare aluminum plate was cleaned with ethyl alcohol and deionized water (DI-water). Then, the base-coat spray was applied on the bare aluminum plate and dried for 1 min. This procedure was repeated three times and the surface was dried for 30 min to obtain a well-coated adhesive layer. Next, the top-coat spray was applied on top of the adhesive layer and dried for 1 min. This process was repeated several times and the surface was dried for 12 h to fix the hydrophobic layer onto the adhesive layer. For the sprayed-on TENG cap, cleaning process with ethyl alcohol and deionized water was omitted.

2.2. Contact Angle Measurement Methods

The contact angle of each surface was measured with a drop-shape analysis device (SmartDrop, Femtofab Co., Seongnam, Korea); the average was determined from five measurements at different locations for each surface.

3. Results and Discussion

The sprayed-on TENG is composed of three layers: an aluminum layer as the base substrate, an adhesive layer in the middle and a hydrophobic layer on the top (Figure 1a). The field-emission scanning electron microscopy (FE-SEM) image on the right shows the top view of the sprayed hydrophobic layer, in which the scale length is 30 µm. As shown in the FE-SEM image, the polymer layer is well-established after a simple spray coating process. Figure 1b shows the entire fabrication process using a commercial hydrophobic spray (NeverWet, RUST-OLEUM). The magnified images below the fabrication schematics are FE-SEM images taken from samples during each process. As shown in Figure 1a, the aluminum surface does not have any micro/nano structures on the surface. As shown in second FE-SEM image (Figure 1b), a sticky polymer layer was formed on the surface of the aluminum substrate. In the third FE-SEM image (Figure 1b, with a scale of 600 nm), hydrophobic nanoparticles were spread and fixed on the adhesive layer and the actual surface area of the top surface increased. According to Wenzel's equations [19], the roughness factor (the ratio between the actual surface and the geometric surface) increases as nanostructures are fabricated on the aluminum substrate and the surface becomes superhydrophobic. Additional FE-SEM images are provided in Supplementary Materials Figure S1, where Figure S1a–d represent the bare aluminum surface, the adhesive layer surface, the hydrophobic surface with the top coat and the magnified hydrophobic surface with the top coat, respectively.

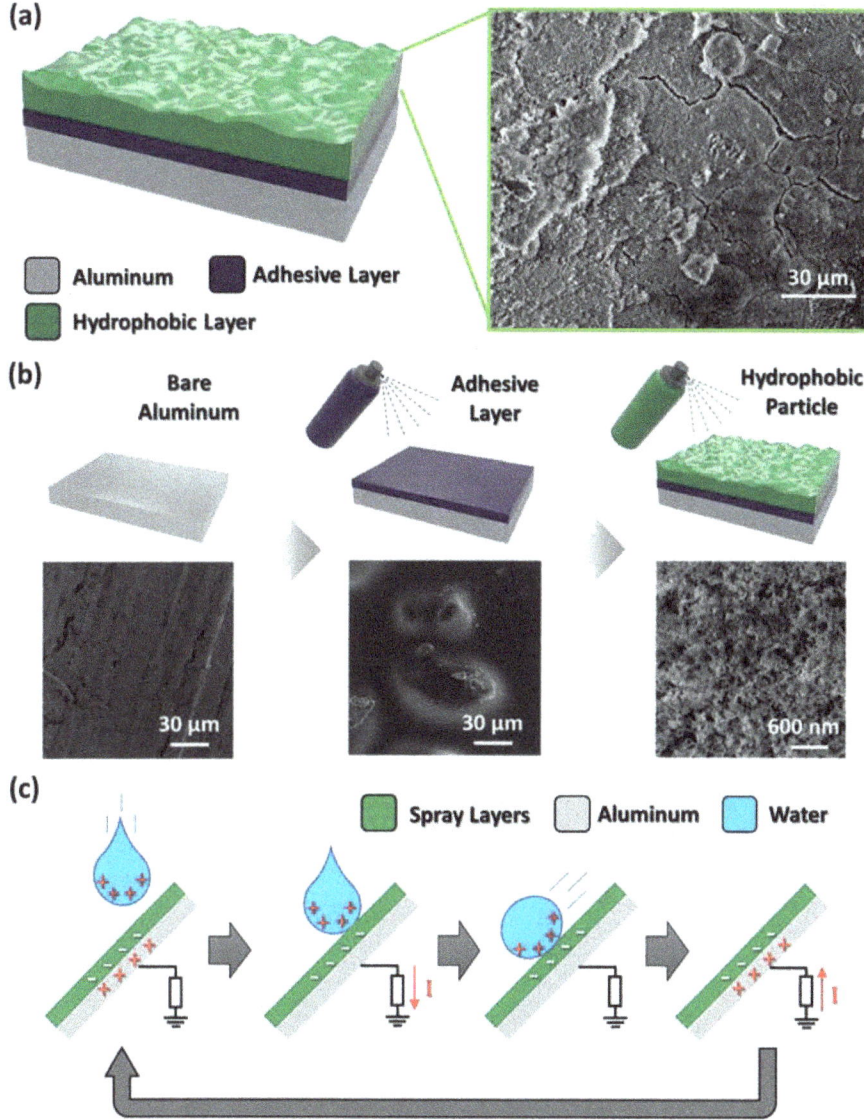

Figure 1. (a) Schematic illustration of sprayed-on superhydrophobic surface and field-emission scanning electron microscopy (FE-SEM) image; (b) Fabrication method of sprayed-on superhydrophobic surface and corresponding FE-SEM images; (c) Working mechanism of sprayed-on TENG.

The working mechanism of the sprayed-on TENG is based on the single-electrode-mode TENG, a liquid-solid contact TENG studied in previous works [14,20]. In Figure 1c, the water droplet is positively pre-charged due to various environmental factors such as friction with air or pipe [13].

And sprayed superhydrophobic surface is negatively pre-charged because water is preliminarily dropped several times and accordingly has triboelectric interactions with the surface. When the water droplet falls onto the sprayed superhydrophobic layer, an electric potential equilibrium is formed, causing current to flow instantaneously from the electrode (aluminum) to the ground. When the water droplet is in complete contact with the superhydrophobic surface, the electrode becomes neutral and no current flows. When the water droplet naturally slides down due to gravity, current reflows instantly from the ground to aluminum layer due to electrostatic induction by the charged hydrophobic layer. This working process is repeated for each water droplet that falls onto the superhydrophobic surface.

Figure 2a shows the contact angle of the sprayed superhydrophobic surface as a function of the number of coats. As shown in the plot, the bare aluminum used in this study showed an average contact angle of 103.2° when 3 µL of DI-water was dropped on the surface. After the first layer of the top coat was applied, the average contact angle increased to 160°. As the number of top coats increased, the contact angle increased slightly.

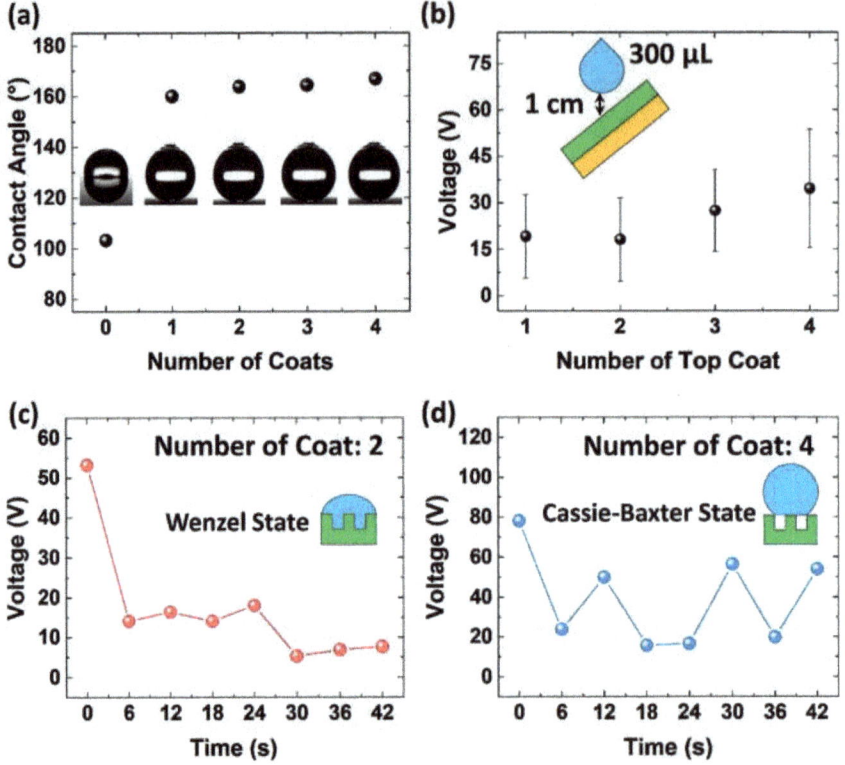

Figure 2. (a) Contact angle and (b) V_{OC} output depending on the number of top coats applied. Peak V_{OC} value of sprayed-on triboelectric nanogenerator (TENG) with (c) 2 top coats and (d) 4 top coats.

However, as shown in Figure 2b, the V_{OC} output of the sprayed-on TENG changed dramatically depending on the number of top coats applied on the aluminum surface. The V_{OC} output was measured when 300 µL of tap water was dropped (1 drop every 6 s) from a height of 1 cm onto the hydrophobic surface. The sprayed-on TENG was tilted approximately 60° to the ground for the water drops to be separated naturally after coming into contact with the TENG surface. With a single top coat, the average V_{OC} output was 19.1 V, which increases to 34.6 V when 4 coats were applied to the TENG surface. This is because the ability of the sprayed hydrophobic surface to withstand the hydraulic pressure of the water drops varies depending on the number of top coats. For a super-hydrophobic surface to be sustained drop after drop, air pockets in between micro-/nanostructures are necessary; this is to ensure that the water drops are in the Cassie-Baxter state. A liquid-solid contact is in the Cassie-Baxter state when two criteria are met: (i) the perimeter of the surface structures are greater than the body forces and (ii) the surface structures are taller than the liquid protruding between them, so that the liquid does not come into contact with the base of the solid [21]. The contact angle of the sprayed superhydrophobic surface using 3 µL of DI-water may show super-hydrophobicity (Figure 2a) but as the hydraulic pressure is increased when using 300 µL of tap water falling from certain height, the liquid can penetrate through the surface structures and come into contact with the base surface. This causes the air pockets in between the micro-/nanostructures to be filled with water (Wenzel state) and the surface is no longer superhydrophobic.

The experimental result is shown in Supplementary Material Figure S2, where the water drops remained on the sprayed hydrophobic surface with 1~2 top coats. The sprayed hydrophobic surface with 3~4 top coats remained dry even after 300 µL of tap water was applied onto the surface. The V_{OC} output of each surface shows same results as well. Figure 2c,d are plots of the peak V_{OC} values when 300 µL of tap water was applied, with 1 drop every 6 s. As shown in Figure 2c, the initial voltage with 2 top coats was 53.2 V, which then decreased drastically. When the first drop of water falls onto the surface, the surface is completely dry, so the sprayed-on TENG can produce a high output. However, the first drop forms a Wenzel state with the solid surface and the water drop is pinned onto the surface. The water remaining from the first drop interferes with the second drop and decreases the electric potential between this second drop and the solid surface [20]. As a result, the V_{OC} output of the sprayed-on TENG decreases after the first drop. In contrast, the water drop forms a Cassie-Baxter state with the superhydrophobic surface with 4 top coats; thus, the water drop rolls off immediately after coming into contact with the solid surface, leaving only a small amount of or no water residue (Figure 2d). As shown in the plot, the peak V_{OC} output decreases after the first drop, similar to the sprayed-on TENG with 2 coats. However, the remaining water residue is small, so it can evaporate or become detached easily from the surface; this results in less interference between the next drop and the solid surface. The experimental results in Figure 2d also shows recovered V_{OC} output during the experiment. Therefore, on the average, the sprayed-on TENG with 4 top coats has a higher output than the sprayed-on TENG with 2 top coats.

Figure 3a shows the continuous V_{OC} output of the sprayed-on TENG with 4 top coats when tap water was sprayed continuously with a shower head. The average positive peak V_{OC} output is 13.4 V. The continuous I_{CC} output is shown in Supplementary Material Figure S3, where the average positive peak I_{CC} output is measured to be 2.1 µA. Also, each corresponding magnified voltage and current graph is shown in Supplementary Material Figure S4. As the lifetime of a sprayed superhydrophobic surface is important, the sprayed-on TENG with 4 top coats was exposed to a continuous water-drop condition using tap water (Figure 3b). Single commercial tap water was dropped every 0.5 s for 20 h. As shown in the plot, the sprayed-on TENG initially produced a peak V_{OC} of about 12 V but after 20 h, the V_{OC} output decreased to 7 V. However, the decreased output was due to the water residue on the superhydrophobic surface. After drying for 3 h, the sprayed-on TENG showed the same V_{OC} value as the initial output (Figure 3c). As shown in Figure 3d, the surfaces before and after 20 h of water application showed no differences when observed with the naked eye; in addition, the contact angle of the superhydrophobic surface remained the same (over 160°). This result indicates that surface structure of the sprayed-on TENG remains the same after the experiment.

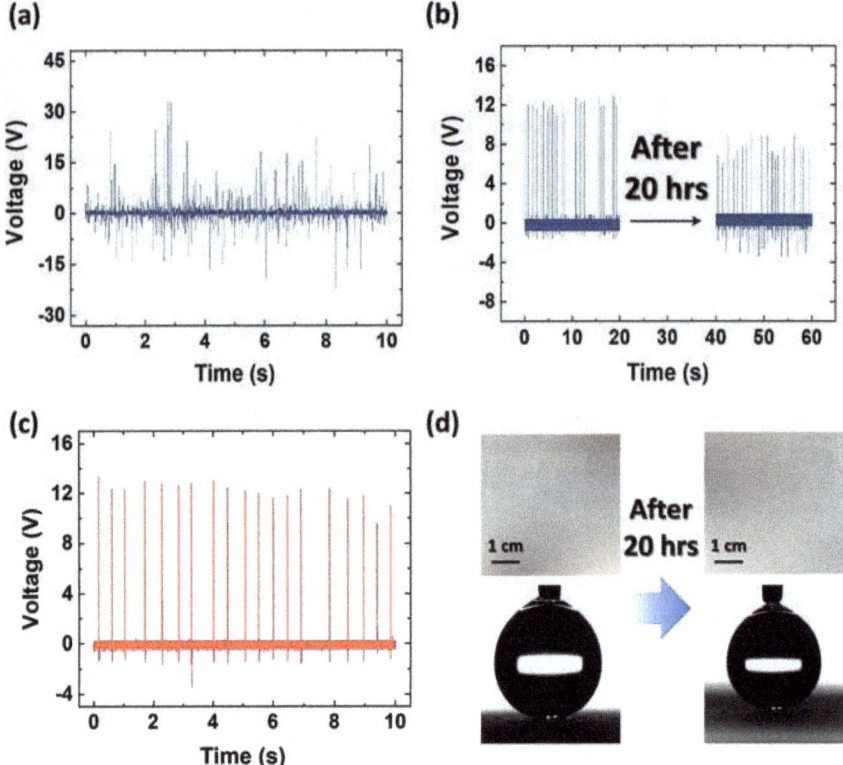

Figure 3. (a) V_{OC} output of sprayed-on TENG with water applied using a commercial shower head; (b) V_{OC} output of sprayed-on TENG after 20 h of water application; (c) Recovered V_{OC} output of sprayed-on TENG after drying; (d) Surface and contact angle of sprayed-on TENG before and after 20 h of water application.

With a spray-on method, a liquid-solid contact TENG can be created on any complex 3D structure. In this work, a liquid-solid contact TENG was fabricated on the brim of a cap, which has a complicated surface structure due to the knitted yarns. Overall, the fabrication process is the same as the process used for the sprayed-on TENG in Figure 1b. Aluminum foil (electrode) was attached to the brim of the cap instead of an aluminum plate. In the sprayed-on TENG cap, the aluminum foil on the brim and all other parts of the cap was coated. Even though only the aluminum foil works as an electrode, both the cap fabric and the aluminum electrode need to be superhydrophobic; this is because if the cap fabric is even partially wet, water molecules can propagate through the yarns and a large area of the fabric gets wet eventually. If the fabric directly in contact with the aluminum foil gets wet, the electrical potential between the falling water drop and the aluminum electrode is significantly reduced, resulting in less or no electrical output. Figure 4a shows an actual photograph of a sprayed-on TENG cap, which is similar to a single-electrode-mode TENG. As shown in the photograph, both the aluminum foil and the cap fabric were coated with the hydrophobic spray.

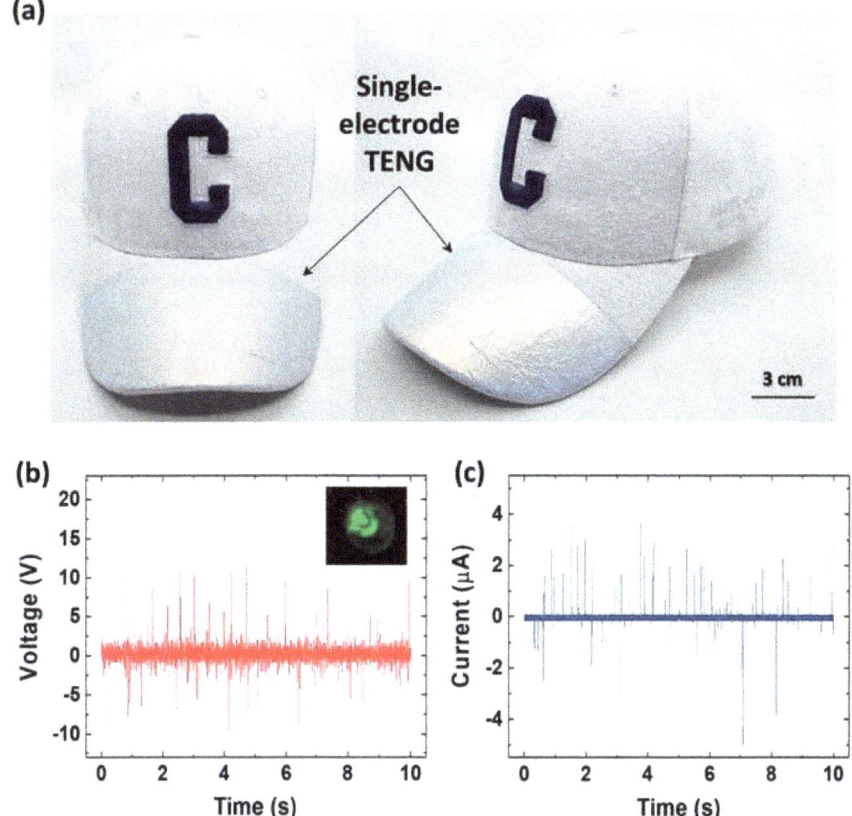

Figure 4. (a) Photograph of sprayed-on TENG cap. (b) V_{OC} and (c) I_{CC} output of sprayed-on TENG cap when water was applied using a commercial shower head.

The fabricated sprayed-on TENG cap was pre-tested and checked for super-hydrophobicity by using tap water sprinkled with a commercial shower head (Supplementary Material Figure S5, Supplementary Video S1). As shown in the photographs, the sprayed-on TENG cap repelled all the water drops from the shower head and there were no water drops left on the surface. Figure 4b,c represent the continuous V_{OC} and I_{CC} outputs of the sprayed-on TENG cap, respectively, during the experiment. The electrical output of the sprayed-on TENG cap was sufficient to light up a LED when connected to a rectifier circuit, as shown in the inset of Figure 4b. Detailed circuit is shown in Supplementary Material Figure S6. In addition, to show a stability of sprayed superhydrophobic coating visually, bending test was conducted on paper, polyimide, polyurethane film sample. First, each 2 cm × 5 cm sample followed same fabrication process used for the sprayed-on TENG in Figure 1b and was checked for super-hydrophobicity by DI-water droplets (Supplementary Video S2). Next, both ends of each sample (about 1 cm) was attached on plates of commercial vibration tester (ET-126B-4, Labworks Co., Costa Mesa, CA, USA) and bending was applied two thousand times by constant amplitude (about 1 cm) of vibration, as shown in Supplementary Video S3. Finally, each sample was successfully checked for super-hydrophobicity by DI-water (Supplementary Video S4). This result demonstrated sprayed superhydrophobic coating has infinite potential to be utilized as TENG by applying all kinds of materials.

4. Conclusions

In summary, we developed a sprayed-on TENG using a commercial hydrophobic spray that can easily create a superhydrophobic surface. The surface has a contact angle of over 160°, which was achieved with only a few spraying processes. The electrical output depends on the number of top coats applied on the solid surface; it was determined that the electrical output could be maximized by maintaining a Cassie-Baxter state between the water drop and the superhydrophobic surface. The sprayed-on superhydrophobic surface produced an average positive peak V_{OC} of 13.4 V and I_{CC} of 2.1 µA under continuous water sprinkling from a commercial shower head. The sprayed-on superhydrophobic surface was able to withstand 20 h of water drops falling every 0.5 s from a tap without surface damage. To demonstrate the easy application of the spray-on method on a complex 3D surface, a superhydrophobic surface was created on the brim of a cap. The sprayed-on TENG cap was able to light up a LED when water was applied. Therefore, this simple spray-on method to create a superhydrophobic surface can be a potential solution for mass-production and commercialization of liquid-solid contact TENGs in the future.

Supplementary Materials: The following are available online at http://www.mdpi.com/2072-666X/9/11/593/s1, Figure S1: Additional FE-SEM Images, Figure S2: Photographs of superhydrophobic surfaces after dropping 300 µL of water, Figure S3: I_{CC} output of sprayed-on TENG when water was applied with a commercial shower head, Figure S4: Magnified V_{OC} and I_{CC} output of sprayed-on TENG when water was applied with a commercial shower head, Figure S5: Photographs of sprayed-on TENG cap when water is applied with a commercial shower head, Figure S6: Sprayed-on TENG connected to a circuit to light an LED, Video S1: Water applied onto the sprayed-on TENG cap when with a commercial shower head, Video S2: DI-water applied onto the paper, polyimide, polyurethane film sample before bending test, Video S3: Bending test with a commercial vibration tester, Video S4: DI-water applied onto the paper, polyimide, polyurethane film sample after bending test.

Author Contributions: J.C. and D.H. contributed equally to this work. J.C. and D.H. performed the experiments and wrote the manuscript. B.K. took the FE-SEM images and analyzed the images. S.L. analyzed the experimental results and reviewed & edited the manuscript.

Funding: This work was supported by the Chung-Ang University Research Scholarship Grant in 2016 and the National Research Foundation of Korea (NRF) grant funded by the Korea government (MSIT) (No. 2016R1A4A1012950).

Conflicts of Interest: The authors declare no conflict of interest. The funders had no role in the design of the study; in the collection, analyses, or interpretation of data; in the writing of the manuscript, or in the decision to publish the results.

References

1. Gibson, T.L.; Kelly, N.A. Solar photovoltaic charging of lithium-ion batteries. *J. Power Sources* **2010**, *195*, 3928–3932. [CrossRef]
2. Cuadras, A.; Gasulla, M.; Ferrari, V. Thermal energy harvesting through pyroelectricity. *Sens. Actuators A Phys.* **2010**, *158*, 132–139. [CrossRef]
3. Jabbar, H.; Song, Y.S.; Jeong, T.T. RF energy harvesting system and circuits for charging of mobile devices. *IEEE Trans. Consum. Electron.* **2010**, *56*, 247–253. [CrossRef]
4. Lee, S.; Bae, S.H.; Lin, L.; Yang, Y.; Park, C.; Kim, S.W.; Cha, S.N.; Kim, H.; Park, Y.J.; Wang, Z.L. Super-flexible nanogenerator for energy harvesting from gentle wind and as an active deformation sensor. *Adv. Funct. Mater.* **2013**, *23*, 2445–2449. [CrossRef]
5. Kim, Y.; Lee, S.; Cho, H.; Park, B.; Kim, D.; Hwang, W. Robust superhydrophilic/hydrophobic surface based on self-aggregated Al2O3 nanowires by single-step anodization and self-assembly method. *ACS Appl. Mater. Interfaces* **2012**, *4*, 5074–5078. [CrossRef] [PubMed]
6. Yen, B.C.; Lang, J.H. A variable-capacitance vibration-to-electric energy harvester. *IEEE Trans. Circuits Syst. I* **2006**, *53*, 288–295. [CrossRef]
7. Yang, B.; Lee, C.; Xiang, W.; Xie, J.; He, J.H.; Kotlanka, R.K.; Low, S.P.; Feng, H. Electromagnetic energy harvesting from vibrations of multiple frequencies. *J. Micromech. Microeng.* **2009**, *19*, 035001. [CrossRef]
8. Fan, F.R.; Tian, Z.Q.; Wang, Z.L. Flexible triboelectric generator! *Nano Energy* **2012**, *1*, 328–334. [CrossRef]

9. Park, C.; Song, G.; Cho, S.M.; Chung, J.; Lee, Y.; Kim, E.H.; Kim, M.; Lee, S.; Huh, J.; Park, C. Supramolecular-Assembled Nanoporous Film with Switchable Metal Salts for a Triboelectric Nanogenerator. *Adv. Funct. Mater.* **2017**, *27*. [CrossRef]
10. Heo, D.; Kim, T.; Yong, H.; Yoo, K.T.; Lee, S. Sustainable oscillating triboelectric nanogenerator as omnidirectional self-powered impact sensor. *Nano Energy* **2018**, *50*, 1–8. [CrossRef]
11. Chung, J.; Yong, H.; Moon, H.; Quang, V.D.; Choi, S.; Kim, D.; Lee, S. Hand-Driven Gyroscopic Hybrid Nanogenerator for Recharging Portable Devices. *Adv. Sci.* **2018**, 1801054. [CrossRef]
12. Wang, X.; Wen, Z.; Guo, H.Y.; Wu, C.S.; He, X.; Lin, L.; Cao, X.; Wang, Z.L. Fully Packaged Blue Energy Harvester by Hybridizing a Rolling Triboelectric Nanogenerator and an Electromagnetic Generator. *ACS Nano* **2016**, *10*, 11369–11376. [CrossRef] [PubMed]
13. Lin, Z.H.; Cheng, G.; Lee, S.; Pradel, K.C.; Wang, Z.L. Harvesting Water Drop Energy by a Sequential Contact-Electrification and Electrostatic-Induction Process. *Adv. Mater.* **2014**, *26*, 4690–4696. [CrossRef] [PubMed]
14. Ha, J.; Chung, J.; Kim, S.; Kim, J.H.; Shin, S.; Park, J.Y.; Lee, S.; Kim, J.B. Transfer-printable micropatterned fluoropolymer-based triboelectric nanogenerator. *Nano Energy* **2017**, *36*, 126–133. [CrossRef]
15. Lee, J.-W.; Hwang, W.J. Theoretical study of micro/nano roughness effect on water-solid triboelectrification with experimental approach. *Nano Energy* **2018**, *52*, 315–322. [CrossRef]
16. Liu, H.; Feng, L.; Zhai, J.; Jiang, L.; Zhu, D. Reversible wettability of a chemical vapor deposition prepared ZnO film between superhydrophobicity and superhydrophilicity. *Langmuir* **2004**, *20*, 5659–5661. [CrossRef] [PubMed]
17. Owen, M.J.; Smith, P.J. Plasma treatment of polydimethylsiloxane. *J. Adhes. Sci. Technol.* **1994**, *8*, 1063–1075. [CrossRef]
18. Lee, Y.; Cha, S.H.; Kim, Y.-W.; Choi, D.; Sun, J.-Y. Transparent and attachable ionic communicators based on self-cleanable triboelectric nanogenerators. *Nat. Commun.* **2018**, *9*, 1804. [CrossRef] [PubMed]
19. Wenzel, R.N. Resistance of solid surfaces to wetting by water. *Ind. Eng. Chem.* **1936**, *28*, 988–994. [CrossRef]
20. Lee, S.; Chung, J.; Kim, D.Y.; Jung, J.Y.; Lee, S.H.; Lee, S. Cylindrical Water Triboelectric Nanogenerator via Controlling Geometrical Shape of Anodized Aluminum for Enhanced Electrostatic Induction. *ACS Appl. Mater. Interfaces* **2016**, *8*, 25014–25018. [CrossRef] [PubMed]
21. Extrand, C. Criteria for ultralyophobic surfaces. *Langmuir* **2004**, *20*, 5013–5018. [CrossRef] [PubMed]

© 2018 by the authors. Licensee MDPI, Basel, Switzerland. This article is an open access article distributed under the terms and conditions of the Creative Commons Attribution (CC BY) license (http://creativecommons.org/licenses/by/4.0/).

Article

A Spherical Hybrid Triboelectric Nanogenerator for Enhanced Water Wave Energy Harvesting

Kwangseok Lee [1], Jeong-won Lee [1], Kihwan Kim [1], Donghyeon Yoo [1], Dong Sung Kim [1], Woonbong Hwang [1,*], Insang Song [2] and Jae-Yoon Sim [3]

1. Department of Mechanical Engineering, Pohang University of Science and Technology (POSTECH), Pohang 37673, Korea; shepherd@postech.ac.kr (K.L.); aaron@postech.ac.kr (J.-w.L.); kihwan@postech.ac.kr (K.K.); ta2two@postech.ac.kr (D.Y.); smkds@postech.ac.kr (D.S.K.)
2. Agency for Defense Development (ADD), Daejeon 34186, Korea; energysong@add.re.kr
3. Department of Electrical Engineering, Pohang University of Science and Technology (POSTECH), Pohang 37673, Korea; jysim@postech.ac.kr
* Correspondence: whwang@postech.ac.kr; Tel.: +82-54-279-2174

Received: 22 October 2018; Accepted: 12 November 2018; Published: 15 November 2018

Abstract: Water waves are a continuously generated renewable source of energy. However, their random motion and low frequency pose significant challenges for harvesting their energy. Herein, we propose a spherical hybrid triboelectric nanogenerator (SH-TENG) that efficiently harvests the energy of low frequency, random water waves. The SH-TENG converts the kinetic energy of the water wave into solid–solid and solid–liquid triboelectric energy simultaneously using a single electrode. The electrical output of the SH-TENG for six degrees of freedom of motion in water was investigated. Further, in order to demonstrate hybrid energy harvesting from multiple energy sources using a single electrode on the SH-TENG, the charging performance of a capacitor was evaluated. The experimental results indicate that SH-TENGs have great potential for use in self-powered environmental monitoring systems that monitor factors such as water temperature, water wave height, and pollution levels in oceans.

Keywords: triboelectric nanogenerator; energy harvesting; hybrid energy; water wave energy

1. Introduction

The demand for energy has steadily increased since the development of the steam engine. In particular, in pursuit of convenience and advancement, the rapid growth and development of industries has prompted humans to explore new energy resources to meet their increasing energy demands. However, considering the need for alternative sources of energy owing to the non-renewable nature of fossil fuels, increasing environmental pollution due to the use of these traditional sources of energy, and consequent global warming, energy harvesting from natural sources (renewable energy) has become a hot topic in the field of energy technology [1–11]. Water waves are a renewable energy source, and water wave energy is continuously generated in oceans all around the world. It is theoretically estimated that 2 TW of energy per year could potentially be generated from water waves, which is more than our current and near-future energy requirements [12]. Although several previous studies have proposed various methods of harvesting water wave energy, the methods proposed in most of those studies are inefficient, complicated, and expensive because they typically involve converting wave energy into translational or rotational energy to operate an electromagnetic generator (EMG) [13–15]. In addition, the random motion and low frequency of water waves are significant hurdles in the generation of power using EMGs [16–18].

Triboelectric nanogenerators (TENGs) can overcome the abovementioned drawbacks as TENGs convert the mechanical energy of water waves into electricity based on a combination of the contact electrification and electrostatic induction phenomena [19,20]. The advantages of energy generation systems developed using TENGs include simple design, low weight, low cost, and high efficiency [19,21–41]. In addition, TENGs can generate energy from solid–liquid as well as solid–solid triboelectrification [42,43]. Because of these advantages, several types of TENGs have been proposed for harvesting water wave energy [44–47]. For example, turbine-based TENGs that generate energy by converting water wave energy into translational or rotational energy have been proposed [48,49]. In addition, encapsulated designs, including those for movable objects, have been proposed for efficiently harvesting the complex motion of water waves [50–53]. Furthermore, to harvest water wave energy via water contact electrification, TENGs with hydrophobic surfaces have been proposed [42,54].

In this work, based on these previous studies on TENGs for water wave energy harvesting, we propose an improved TENG called the spherical hybrid triboelectric nanogenerator (SH-TENG). By harvesting energy from multiple sources through one electrode simultaneously, our proposed SH-TENG provides a simplified approach to previous hybrid TENG systems [54–56]. Moreover, the simple fabrication process used for the proposed SH-TENG easily facilitates mass production, which can contribute to the commercialization and scalability of the proposed SH-TENG. Thus, SH-TENGs have great potential for use in self-powered environmental monitoring systems that monitor factors such as water temperature, water wave height, and pollution levels in oceans.

2. Experimental Section

2.1. Materials and Methods

The proposed SH-TENG was fabricated as follows. Industrial aluminum hemispherical shells with a diameter of 6 cm were immersed in 1 M NaOH for 5 min and then cleaned using deionized (D.I.) water. The cleaned hemispherical shells were masked in the center of the inner surface with a maskant (HTM-201, Woo Ju Chemical Co., Incheon-si, Korea) and a wire was connected to the edge of the shells. Then, the hemispherical shells were anodized in a 0.3 M oxalic acid solution at a constant voltage of 50 V for 100 min using a power supply (DRP-92001DUS, Digital Electronics Co., Incheon-si, Korea). The temperature of the solution was constantly maintained using a circulator (RW-0525G, Lab Companion Inc., Seoul-si, Korea) during the anodization process. Thereafter, the as-prepared shells were dipped in a mixture of n-hexane and heptadecafluoro-1,1,2,2,-tetrahydrodecyl trichlorosilane ($C_{10}H_4Cl_3F_{17}Si$, HDFS, Alfa Aesar, MA, USA), with a volumetric ratio of 1000:1, for 10 min and dried in an oven at 110 °C for 10 min. The coated hemispherical shells were then soaked in a solvent ($C_{10}HF_{22}N$, FC-40, Fluorinert™, Saint Paul, MN, USA) with 0.1 wt% polytetrafluoroethylene (($C_2F_4)_n$, PTFE, Chemours™, Wilmington, DE, USA), after which they were dried in an oven under vacuum at 60 °C for 24 h. The edge of the re-coated hemispherical shells was then covered with 1 mm thick polyamide adhesive to isolate each shell (Figure S1). Finally, a 3 cm diameter nylon ball was placed inside one shell and the two shells were joined together (Figure 1).

2.2. Characterization

The surface structure and cross section of the fabricated SH-TENG were observed using a field emission scanning electron microscope (SU66000, Hitachi, Tokyo, Japan). To evaluate the wettability of the SH-TENG surface, contact angles (CAs) were measured using a D.I. water droplet analysis tool (SmartDrop, FEMTOFAB, Seongnam-si, Korea). The D.I. water droplet was 5 µL.

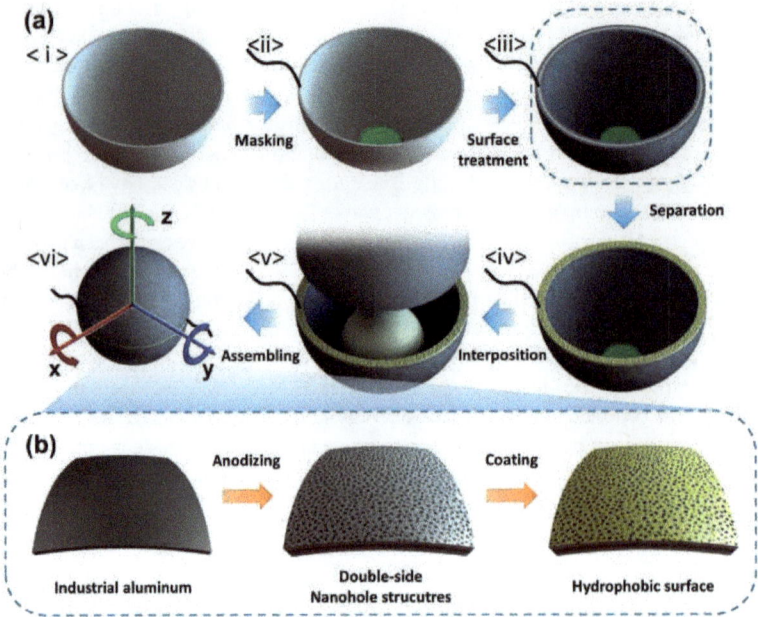

Figure 1. (**a**) Fabrication of the spherical hybrid triboelectric nanogenerator (SH-TENG). (**b**) Surface treatment of hydrophobic nanohole-structured anodic aluminum oxide.

2.3. Electrical Output Measurement of the Spherical Hybrid Triboelectric Nanogenerator (SH-TENG)

To analyze the electrical outputs from both the inner and outer surfaces of the SH-TENG, solid–solid and solid–liquid TENGs were fabricated in addition to the previously specified SH-TENG. More specifically, the solid–solid TENG was fabricated by covering the outer surface of the SH-TENG to prevent solid–liquid triboelectrification at the outer surface; in contrast, the solid–liquid TENG was fabricated by removing the nylon ball from the SH-TENG to prevent solid–solid triboelectrification at the inner surface. The electrical output of these three TENGs was then analyzed in a motion plan involving six degrees of freedom (6 DoFs), which were used to represent both a change in position and rotation along a perpendicular axis (Figure S2). The output voltage of the SH-TENG for translational and rotational motion was measured using a mixed signal-and-digital oscilloscope (DS1074Z, RIGOL, Beaverton, OR, USA). To vibrate the SH-TENG, it was mounted on an electrodynamic shaker (ET-140, Labworks Inc., Costa Mesa, CA, USA). In particular, for vertical vibrations, the SH-TENG was set to sink 2 cm in D.I. water and then rise with a frequency of 1 Hz. For horizontal vibrations, the SH-TENG was vibrated horizontally with 2 cm of it submerged in the water at a frequency of 1 Hz. Furthermore, to evaluate the output of the SH-TENG during rotational movement, with 2 cm of the SH-TENG submerged in D.I. water, it was rotated using a speed-control motor (E9I90PBH-TU, ExceM, Seoul-si, Korea); the rotational velocity was fixed at 60 rpm. In order to observe the electrical output of these TENGs during water wave motion, they were placed in a 100 cm × 70 cm × 20 cm bath. Then, a controllable water pump (EcoDrift 15.0, Aquamedic, Bissendorf, Germany) was operated to generate water waves. To rectify the electrical output, a rectifier (B40C800DM-E3/45, Vishay Semiconductors, Malvern, PA, USA) was used.

3. Results and Discussion

3.1. Design and Features of the SH-TENG

A symmetric spherical design was adopted for our proposed SH-TENG in order to reduce loss of kinetic energy and enable flexible movement of the SH-TENG under the effect of random water waves. As shown in Figure 2a, all surfaces of the SH-TENG undergo water contact electrification because the hemispherical shell is made of aluminum, which acts as an electrode. In particular, the hemispherical shell is covered on both the inner and outer surfaces with a dual layer consisting of Al_2O_3 and PTFE. The Al_2O_3 layer serves as a dielectric layer that acts as an insulator to prevent leakage of charge (Figure 2d) [57]. PTFE, which has a strong natural tendency to be negatively charged in triboelectric series, is used as the friction layer in the SH-TENG because water tends to acquire a positive charge on triboelectrification [58]. Thus far, several TENGs have been manufactured that involve bonding or depositing these different layers separately [48–56]. However, in the case of the proposed SH-TENG, the bonding process is omitted because anodizing and self-assembled monolayer (SAM) coating methods are used to robustly develop Al_2O_3 and PTFE layers on the electrode without the need for adhesives; this is an advantage for mass production of the proposed SH-TENG. Furthermore, a nanohole structure is applied to the surface of the SH-TENG to enable the separation of water from the surface after contact, as well as to simultaneously maximize the frictional force experienced by the SH-TENG; the nanostructure is shown in Figure 2c [59]. In order to prevent the two shells from making electrical contact, a hot melt glue stick was used to cover the edge of the shells with a 1 mm thick polyamide layer. The high viscosity of the molten polyamide enabled it to be applied thickly to electrically isolate each shell. As previously mentioned, the hemispherical shells were partially masked with HTM-201 to prevent contact electrification in that area. Further, this masking enables the nylon ball to separate from the PTFE after making contact inside the hemispherical shell.

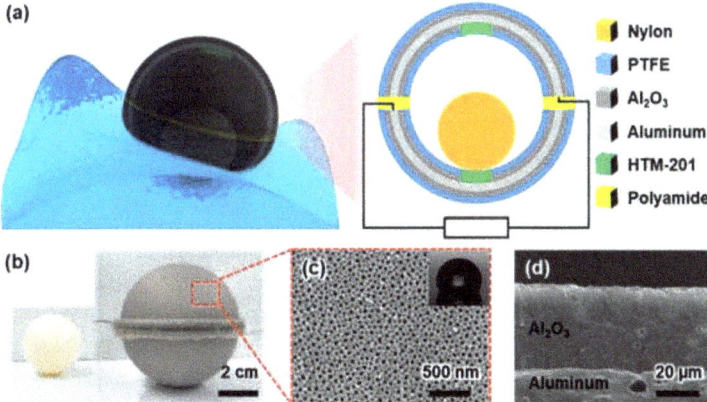

Figure 2. (a) Three-dimensional model and schematic diagram of the SH-TENG. (b) Photograph of the nylon ball and SH-TENG. (c) Scanning electron microscopy (SEM) image and contact angles (CAs) of the SH-TENG surface. (d) Cross-sectional SEM image of Al_2O_3.

3.2. Operating Principle

The proposed SH-TENG should be able to realize both solid–solid triboelectrification—which occurs as a result of contact separation between the PTFE surface and the nylon ball inside the SH-TENG—and solid–liquid triboelectrification, simultaneously. Figure 3(a1) shows a schematic diagram depicting triboelectrification between the hemispherical shell and the nylon ball. When the PTFE surface is in contact with the nylon ball, it has a negative charge, while the nylon ball has a positive charge. However, when the ball moves to the masked area owing to translational or rotational motion, electrostatic induction occurs because of the separation of the nylon ball from the PTFE. Then, when the nylon ball again comes into contact with the PTFE layer, contact electrification again occurs. Thus, as long as the SH-TENG is in motion, there will be repeated contact electrification and electrostatic induction. This series of processes can be considered as a single-electrode mode owing to contact separation of the electrified body for a single electrode [60]. Figure 3(a2) shows the triboelectric relationship between the nylon ball and two hemispherical shells. The electrons travel between the two hemispherical shells owing to the potential difference generated by contact electrification. Thus, a current is generated. This design, in which one dielectric object travels between two electrode layers, is similar to that of the freestanding triboelectric mode [61].

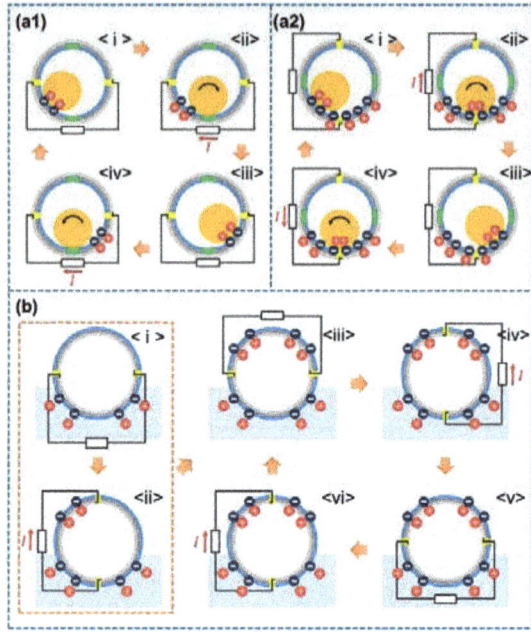

Figure 3. Schematic diagram representing the operating principle of the SH-TENG on its inner surface: (a1) single electrode triboelectric mode, and (a2) freestanding triboelectric mode. (b) Schematic diagram of the operating principle of the SH-TENG on its outer surface.

On the outer surface of the SH-TENG, solid–liquid triboelectrification between the PTFE layer and water leads to energy generation as well. When the SH-TENG is dipped in water, the outer surface of the SH-TENG gets negatively charged, while the surrounding water gets positively charged. This charged state of the SH-TENG is illustrated in Figure 3b <i>. As the SH-TENG is rotated by an external force, the immersed PTFE surface gets negatively charged. The PTFE in contact with water maintains an electrical balance with the water it is in contact with; however, the part of the PTFE that is not in contact with water experiences electrostatic induction in order to balance the electric charge generated on the immersed PTFE layer. This process causes charge transfer to occur between the two hemispherical shells. When the difference of the water contact area between the electrodes is ΔA, the electrical output is determined by the changing rate of ΔA (Equation (1)):

$$V \propto \frac{d\Delta A}{dt} = \frac{d}{dt} \text{(difference of area in contact with water in each electrode)} \quad (1)$$

After all the surfaces have been charged once, the process indicated by Figure 3b <iii>–<vi> is repeated.

3.3. Electrical Output of SH-TENG during 6 Degrees of Freedom (DoF) Motion

Solid–solid, solid–liquid, and hybrid electrical outputs were evaluated for 6 DoF motion (Figure 4). In order to effectively represent the different movements for the 6 DoF motion, an intrinsic coordinate system for the SH-TENG—depicted in Figure 1a <vi>—was used. Motion along the x- and y-axes of the SH-TENG is considered to be the same owing to the symmetric design of the SH-TENG. The electrical outputs from the translational and rotational motion about the x- and z-axes of the SH-TENG were measured.

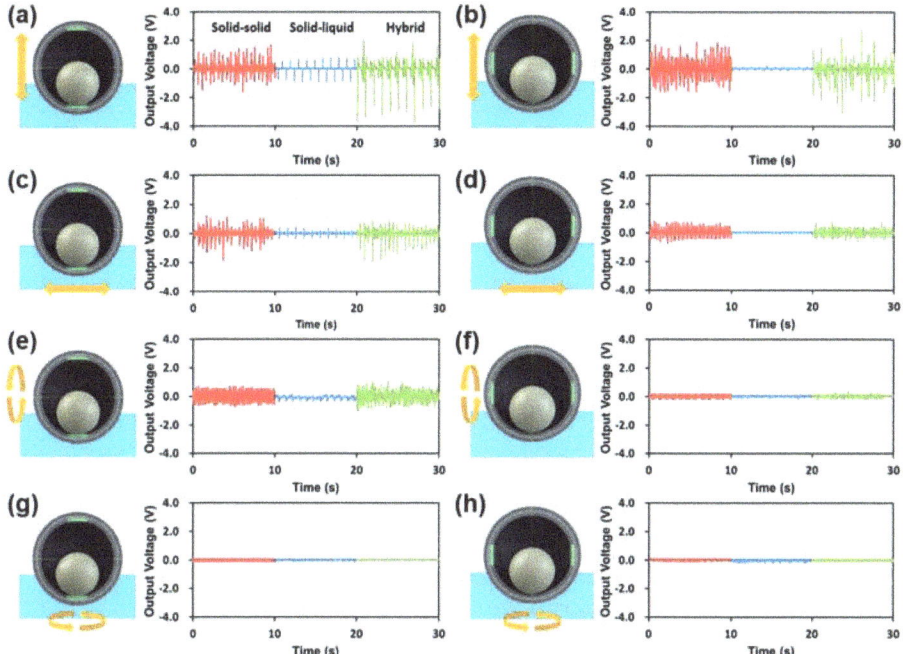

Figure 4. Schematic diagram and electrical output of the triboelectric nanogenerators (TENG): (**a**,**b**) during vertical vibration with respect to the (**a**) z- and (**b**) x-axes; (**c**,**d**) during horizontal vibration with respect to the (**c**) x- and (**d**) z-axes; (**e**,**f**) during rotation with respect to the (**e**) x-axis and (**f**) z-axis parallel to the water surface; (**g**,**h**) during rotation with respect to the (**g**) z-axis and (**h**) x-axis perpendicular to the water surface.

Figure 4a–h includes graphs showing the electrical output voltage of the TENGs during translational or rotational motion. In particular, Figure 4a–d shows the electrical output during translational motion. Figure 4a,b shows the electrical output when the TENG is vertically immersed in D.I. water by 2 cm in the x- and z-axes directions of the SH-TENG. Because of the random motion of the nylon ball inside the SH-TENG, despite constant vibration, a relatively random electrical output is observed for solid–solid triboelectrification. In contrast, for solid–liquid triboelectrification, the electrical output is generated owing to the change in the area in contact with water between the two shells. In the case depicted in Figure 4a, water contact electrification of the lower shell leads to generation of an electrical output because the upper shell is not in contact with water. However, the low electrical output in the case shown in Figure 4b occurs because of the relative lack of difference in the water contact area between each electrode. Figure 4c,d shows graphs depicting the generated electrical output when horizontal vibration is applied to the SH-TENG while it is submerged 2 cm in the water. As in the case with vertical vibration, the randomly distributed energy output is attributed to the random motion of the nylon ball inside the SH-TENG. In particular, because of the small change in the water contact area during solid–liquid triboelectrification, its energy output is considerably lower.

For rotational motion, the energy output was measured with the SH-TENG immersed 2 cm in the water. The observed electrical outputs depend on the contact separation of the nylon ball as well as the change in the contact area between the two hemispherical shells and water. Figure 4e shows the electrical output when the SH-TENG is rotated with respect to the *x*-axis parallel to the water surface. For the other rotation, Figure 4f–h indicates that the electrical output is considerably lower because the nylon ball does not undergo contact separation motion with the PTFE; thus, its contact area with the water does not change.

3.4. Electrical Output of SH-TENG for Water Wave Motion

Figure 5a,b illustrates the actual experimental setup and schematic diagram of the observed electrical output due to the water wave motion. The energy yield of the SH-TENG is evaluated for a 1 Hz water wave in the tank. As shown in Figure 5c, sporadic energy generation was observed per the contact separation phenomenon in the case of solid–solid triboelectrification; the maximum output voltage observed in this case was 2.7 V. In contrast, in the case of solid–liquid triboelectrification, it was confirmed that the energy generation occurred in a continuous form rather than in a sporadic form, because the electrical output is generated due to the change in the contact area between each electrode and the water; the maximum voltage observed in this case was 0.5 V. The combination of these two energies, i.e., energies of the sporadic and continuous form, are referred to as the hybrid form; the maximum peak voltage in this case was 3.3 V.

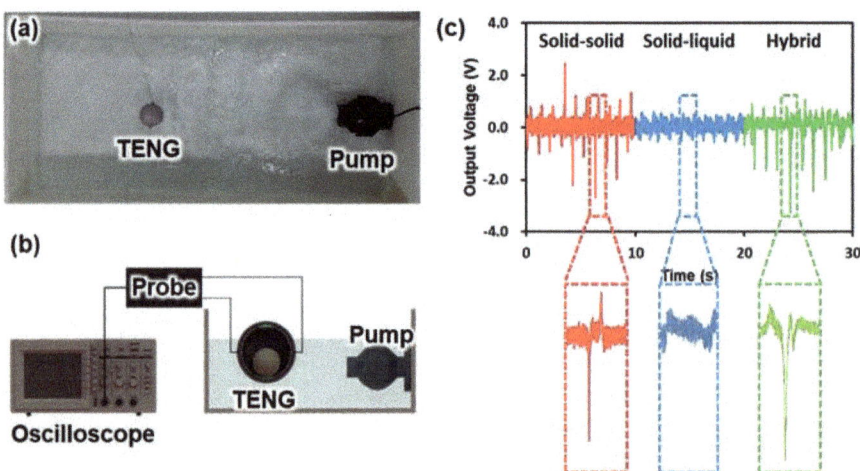

Figure 5. Experimental setup: (**a**) photograph and (**b**) schematic diagram of TENG for electrical output measurements in water wave motion. (**c**) Electrical output of TENG under wave motion with a frequency of 1 Hz.

In addition, to demonstrate whether the energy is harvested by the inner and outer surfaces simultaneously, the circuit shown in Figure 6a was constructed to charge a capacitor using the harvested water wave energy. Figure 6b shows the measured voltage of the 1 µF capacitor charged by the three types of TENGs for 20 s during water wave motion with a wave frequency of 1 Hz. The capacitor was charged to 0.08 V during solid–solid triboelectrification, 0.31 V during solid–liquid triboelectrification, and 0.39 V during hybrid triboelectrification. The maximum voltage during solid–solid triboelectrification is higher than that of solid–liquid triboelectrification as shown in Figure 5c; however, the output of the solid–liquid triboelectrification in terms of capacitor charging is higher than that of solid–solid triboelectrification in the case shown in Figure 6. It can be seen that there is a difference between the maximum peak level and the root mean square (RMS) level. Thus, our experimental results demonstrate that solid–solid triboelectrification and solid–liquid triboelectrification can be simultaneously harvested using one electrode. In addition, the charged energy of the 1 µF capacitor is shown in Figure S3. The charged energy can be derived from the charged voltage in the capacitor by the following equation (Equation (2)):

$$E = \frac{1}{2}CV^2 \qquad (2)$$

where E, C, and V are the charged energy, size of the capacitance, and charged voltage, respectively. The charged energy had a similar tendency to the voltage in Figure 6b by Equation (2) (Figure S3).

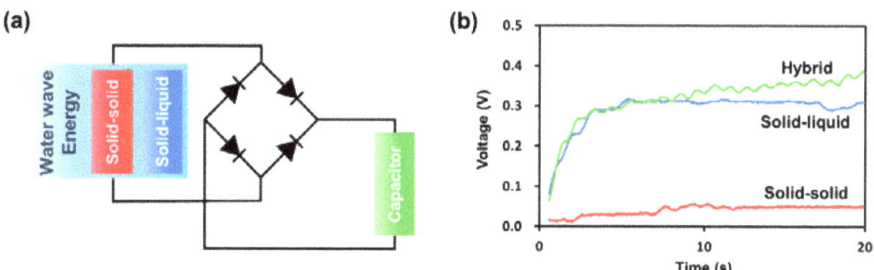

Figure 6. (a) Schematic diagram of the rectifying circuit. (b) Voltage of a 1 µF capacitor charged by the TENG for 20 s during water wave motion with a 1 Hz wave frequency.

4. Conclusions

SH-TENG, which flexibly responds to random water wave energy and harvests hybrid energy using a single electrode, was proposed in this paper for enhanced water wave energy harvesting. The electrical output was analyzed in 6 DoF motion for solid–solid, solid–liquid, and hybrid triboelectrification. In addition, using a capacitor, it was confirmed that solid–solid triboelectrification and solid–liquid triboelectrification are harvested simultaneously through the two sides of one electrode layer. These results indicate that it is possible to simplify the previously proposed hybrid TENG system to enable mass production of TENGs, which can contribute to the commercialization and scalability of blue (i.e., water wave) energy.

Supplementary Materials: The following are available online at http://www.mdpi.com/2072-666X/9/11/598/s1.

Author Contributions: Conceptualization, K.L.; Data curation, K.L.; Formal analysis, K.L.; Funding acquisition, D.S.K., J.-Y.S., and W.H.; Investigation, K.L., J.-w.L., K.K., and D.Y.; Methodology, K.L.; Project administration, J.-w.L. and W.H.; Resources, D.S.K. and W.H.; Supervision, I.S., D.S.K., J.-Y.S., and W.H.; Validation, D.S.K., J.-Y.S., and W.H.; Visualization, K.L. and J.-w.L.; Writing—Original Draft, K.L.; Writing—Review and Editing, J.-w.L. and W.H.

Funding: This research was supported by the Agency for Defense Development in Korea (UE161007RD), and by the National Research Foundation of Korea (NRF) through a grant funded by the Korea Government (MSIT) (grant number NRF-2016K1A3A1A20005945).

Conflicts of Interest: The authors declare no conflict of interest.

References

1. Schiermeier, Q.; Tollefson, J.; Scully, T.; Witze, A.; Morton, O. Electricity without Carbon. *Nature* **2008**, *454*, 816–823. [CrossRef] [PubMed]

2. Glaser, J.A. US renewable energy consumption. *Clean. Techn. Environ. Policy* **2007**, *9*, 249–252. [CrossRef]
3. Clery, D. Renewable energy. U.K. ponders world's biggest tidal power scheme. *Science* **2008**, *320*, 1574. [CrossRef] [PubMed]
4. Ku, M.-L.; Li, W.; Chen, Y.; Ray Liu, K.J. Advances in Energy Harvesting Communications: Past, Present, and Future Challenges. *IEEE Commun. Surv. Tutor.* **2016**, *18*, 1384–1412. [CrossRef]
5. Tian, B.; Kempa, T.J.; Lieber, C.M. Single nanowire photovoltaics. *Chem. Soc. Rev.* **2009**, *38*, 16–24. [CrossRef] [PubMed]
6. Chen, J.; Huang, Y.; Zhang, N.; Zou, H.; Liu, R.; Tao, C.; Fan, X.; Wang, Z.L. Micro-cable structured textile for simultaneously harvesting solar and mechanical energy. *Nat. Energy* **2016**, *1*, 16138. [CrossRef]
7. Wang, Z.L.; Zhu, G.; Yang, Y.; Wang, S.; Pan, C. Progress in nanogenerators for portable electronics. *Mater. Today* **2012**, *15*, 532–543. [CrossRef]
8. Wang, X.; Song, J.; Liu, J.; Wang, Z.L. Direct-Current Nanogenerator Driven by Ultrasonic Waves. *Science* **2007**, *316*, 102–105. [CrossRef] [PubMed]
9. Zhang, N.; Chen, J.; Huang, Y.; Guo, W.; Yang, J.; Du, J.; Fan, X.; Tao, C. A Wearable All-Solid Photovoltaic Textile. *Adv. Mater.* **2016**, *28*, 263–269. [CrossRef] [PubMed]
10. Lin, Z.; Chen, J.; Li, X.; Li, J.; Liu, J.; Awais, Q.; Yang, J. Broadband and three-dimensional vibration energy harvesting by a non-linear magnetoelectric generator. *Appl. Phys. Lett.* **2016**, *109*, 253903. [CrossRef]
11. Song, J.; Wang, Z.L. Piezoelectric Nanogenerators Based on Zinc Oxide Nanowire Arrays. *Science* **2006**, *312*, 242–246. [CrossRef]
12. Ellabban, O.; Abu-Rub, H.; Blaabjerg, F. Renewable energy resources: Current status, future prospects and their enabling technology. *Renew. Sustain. Energy Rev.* **2014**, *39*, 748–764. [CrossRef]
13. Wen, Z.; Guo, H.; Zi, Y.; Yeh, M.H.; Wang, X.; Deng, J.; Wang, J.; Li, S.; Hu, C.; Zhu, L.; et al. Harvesting Broad Frequency Band Blue Energy by a Triboelectric-Electromagnetic Hybrid Nanogenerator. *ACS Nano* **2016**, *10*, 6526–6534. [CrossRef] [PubMed]
14. Callaway, E. Energy: To Catch a Wave. *Nature* **2007**, *450*, 156–159. [CrossRef] [PubMed]
15. Scruggs, J.; Jacob, P. Engineering. Harvesting Ocean Wave Energy. *Science* **2009**, *323*, 1176–1178. [CrossRef] [PubMed]
16. Forristall, G.Z. Measurements of a saturated range in ocean wave spectra. *J. Geophys. Res.* **1981**, *86*. [CrossRef]
17. Bouws, E.; Günther, H.; Rosenthal, W.; Vincent, C.L. Similarity of the wind wave spectrum in finite depth water: 1. Spectral form. *J. Geophys. Res.-Oceans* **1985**, *90*, 975–986. [CrossRef]
18. Toffoli, A.; Bitner-Gregersen, E.M. Types of Ocean Surface Waves, Wave Classification. *Encycl. Marit. Offshore Eng.* **2017**, 1–8. [CrossRef]
19. Fan, F.-R.; Tian, Z.-Q.; Wang, Z.L. Flexible triboelectric generator. *Nano Energy* **2012**, *1*, 328–334. [CrossRef]
20. Wang, Z.L. On Maxwell's displacement current for energy and sensors: The origin of nanogenerators. *Mater. Today* **2017**, *20*, 74–82. [CrossRef]
21. Wang, Z.L. Triboelectric nanogenerators as new energy technology and self-powered sensors-principles, problems and perspectives. *Faraday Discuss.* **2014**, *176*, 447–458. [CrossRef] [PubMed]
22. Zi, Y.; Guo, H.; Wen, Z.; Yeh, M.H.; Hu, C.; Wang, Z.L. Harvesting Low-Frequency (<5 Hz) Irregular Mechanical Energy: A Possible Killer Application of Triboelectric Nanogenerator. *ACS Nano* **2016**, *10*, 4797–4805. [CrossRef] [PubMed]
23. Wang, S.; Lin, L.; Wang, Z.L. Nanoscale triboelectric-effect-enabled energy conversion for sustainably powering portable electronics. *Nano Lett.* **2012**, *12*, 6339–6346. [CrossRef] [PubMed]
24. Wang, S.; Lin, L.; Xie, Y.; Jing, Q.; Niu, S.; Wang, Z.L. Sliding-triboelectric nanogenerators based on in-plane charge-separation mechanism. *Nano Lett.* **2013**, *13*, 2226–2233. [CrossRef] [PubMed]
25. Wang, S.; Niu, S.; Yang, J.; Lin, L.; Wang, Z.L. Quantitative Measurements of Vibration Amplitude Using a Contact-Mode Freestanding Triboelectric Nanogenerator. *ACS Nano* **2014**, *8*, 12004–12013. [CrossRef] [PubMed]
26. Wang, S.; Xie, Y.; Niu, S.; Lin, L.; Wang, Z.L. Freestanding triboelectric-layer-based nanogenerators for harvesting energy from a moving object or human motion in contact and non-contact modes. *Adv. Mater.* **2014**, *26*, 2818–2824. [CrossRef] [PubMed]
27. Lin, Z.; Yang, J.; Li, X.; Wu, Y.; Wei, W.; Liu, J.; Chen, J.; Yang, J. Large-Scale and Washable Smart Textiles Based on Triboelectric Nanogenerator Arrays for Self-Powered Sleeping Monitoring. *Adv. Func. Mater.* **2018**, *28*, 1704112. [CrossRef]

28. Wang, Z.L.; Chen, J.; Lin, L. Progress in triboelectric nanogenerators as a new energy technology and self-powered sensors. *Energy Environ. Sci.* **2015**, *8*, 2250–2282. [CrossRef]
29. Chen, J.; Wang, Z.L. Reviving Vibration Energy Harvesting and Self-Powered Sensing by a Triboelectric Nanogenerator. *Joule* **2017**, *1*, 480–521. [CrossRef]
30. Zhu, G.; Chen, J.; Zhang, T.; Jing, Q.; Wang, Z.L. Radial-arrayed rotary electrification for high performance triboelectric generator. *Nat. Commun.* **2014**, *5*, 3426. [CrossRef] [PubMed]
31. Chen, J.; Zhu, G.; Yang, W.; Jing, Q.; Bai, P.; Yang, Y.; Hou, T.C.; Wang, Z.L. Harmonic-resonator-based triboelectric nanogenerator as a sustainable power source and a self-powered active vibration sensor. *Adv. Mater.* **2013**, *25*, 6094–6099. [CrossRef] [PubMed]
32. Jin, L.; Chen, J.; Zhang, B.; Deng, W.; Zhang, L.; Zhang, H.; Huang, X.; Zhu, M.; Yang, W.; Wang, Z.L. Self-Powered Safety Helmet Based on Hybridized Nanogenerator for Emergency. *ACS Nano* **2016**, *10*, 7874–7881. [CrossRef] [PubMed]
33. Yang, W.; Chen, J.; Zhu, G.; Yang, J.; Bai, P.; Su, Y.; Jing, Q.; Cao, X.; Wang, Z.L. Harvesting Energy from the Natural Vibration of Human Walking. *ACS Nano* **2013**, *7*, 11317–11324. [CrossRef] [PubMed]
34. Zhu, G.; Bai, P.; Chen, J.; Wang, Z.L. Power-generating shoe insole based on triboelectric nanogenerators for self-powered consumer electronics. *Nano Energy* **2013**, *2*, 688–692. [CrossRef]
35. Hou, T.-C.; Yang, Y.; Zhang, H.; Chen, J.; Chen, L.-J.; Wang, Z.L. Triboelectric nanogenerator built inside shoe insole for harvesting walking energy. *Nano Energy* **2013**, *2*, 856–862. [CrossRef]
36. Lin, Z.; Chen, J.; Li, X.; Zhou, Z.; Meng, K.; Wei, W.; Yang, J.; Wang, Z.L. Triboelectric Nanogenerator Enabled Body Sensor Network for Self-Powered Human Heart-Rate Monitoring. *ACS Nano* **2017**, *11*, 8830–8837. [CrossRef] [PubMed]
37. Bai, P.; Zhu, G.; Lin, Z.-H.; Jing, Q.; Chen, J.; Zhang, G.; Ma, J.; Wang, Z.L. Integrated Multilayered Triboelectric Nanogenerator for Harvesting Biomechanical Energy from Human Motions. *ACS Nano* **2013**, *7*, 3713–3719. [CrossRef] [PubMed]
38. Yang, J.; Chen, J.; Yang, Y.; Zhang, H.; Yang, W.; Bai, P.; Su, Y.; Wang, Z.L. Broadband Vibrational Energy Harvesting Based on a Triboelectric Nanogenerator. *Adv. Energy Mater.* **2014**, *4*, 1301322. [CrossRef]
39. Chen, J.; Zhu, G.; Yang, J.; Jing, Q.; Bai, P.; Yang, W.; Qi, X.; Su, Y.; Wang, Z.L. Personalized Keystroke Dynamics for Self-Powered Human-Machine Interfacing. *ACS Nano* **2015**, *9*, 105–116. [CrossRef] [PubMed]
40. Li, Z.; Chen, J.; Zhou, J.; Zheng, L.; Pradel, K.C.; Fan, X.; Guo, H.; Wen, Z.; Yeh, M.-H.; Yu, C.; et al. High-efficiency ramie fiber degumming and self-powered degumming wastewater treatment using triboelectric nanogenerator. *Nano Energy* **2016**, *22*, 548–557. [CrossRef]
41. Liu, R.; Kuang, X.; Deng, J.; Wang, Y.-C.; Wang, A.C.; Ding, W.; Lai, Y.-C.; Chen, J.; Wang, P.; Lin, Z.; et al. Shape Memory Polymers for Body Motion Energy Harvesting and Self-Powered Mechanosensing. *Adv. Mater.* **2018**, *30*, 1705195. [CrossRef] [PubMed]
42. Lin, Z.H.; Cheng, G.; Lin, L.; Lee, S.; Wang, Z.L. Water-solid surface contact electrification and its use for harvesting liquid-wave energy. *Angew. Chem. Int. Ed.* **2013**, *52*, 12777–12781. [CrossRef]
43. Lin, Z.H.; Cheng, G.; Lee, S.; Pradel, K.C.; Wang, Z.L. Harvesting water drop energy by a sequential contact-electrification and electrostatic-induction process. *Adv. Mater.* **2014**, *26*, 4690–4696. [CrossRef] [PubMed]
44. Hu, Y.; Yang, J.; Jing, Q.; Niu, S.; Wu, W.; Wang, Z.L. Triboelectric Nanogenerator Built on Suspended 3D Spiral Structure as Vibration and Positioning Sensor and Wave Energy Harvester. *ACS Nano* **2013**, *7*, 10424–10432. [CrossRef] [PubMed]
45. Chen, J.; Yang, J.; Li, Z.; Fan, X.; Zi, Y.; Jing, Q.; Guo, H.; Wen, Z.; Pradel, K.C.; Niu, S.; et al. Networks of Triboelectric Nanogenerators for Harvesting Water Wave Energy: A Potential Approach toward Blue Energy. *ACS Nano* **2015**, *9*, 3324–3331. [CrossRef] [PubMed]
46. Jing, Q.; Zhu, G.; Bai, P.; Xie, Y.; Chen, J.; Han, R.P.S.; Wang, Z.L. Case-Encapsulated Triboelectric Nanogenerator for Harvesting Energy from Reciprocating Sliding Motion. *ACS Nano* **2014**, *8*, 3836–3842. [CrossRef] [PubMed]
47. Chen, J.; Yang, J.; Guo, H.; Li, Z.; Zheng, L.; Su, Y.; Wen, Z.; Fan, X.; Wang, Z.L. Automatic Mode Transition Enabled Robust Triboelectric Nanogenerators. *ACS Nano* **2015**, *9*, 12334–12343. [CrossRef] [PubMed]
48. Xie, Y.; Wang, S.; Long, L.; Jing, Q.; Lin, Z.-H.; Niu, S.; Wu, Z.; Wang, Z.L. Rotary Triboelectric Nanogenerator Based on a Hybridized Mechanism for Harvesting Wind Energy. *ACS Nano* **2013**, *7*, 7119–7125. [CrossRef] [PubMed]

49. Cheng, G.; Lin, Z.-H.; Du, Z.; Wang, Z.L. Simultaneously Harvesting Electrostatic and Mechanical Energies from Flowing Water by a Hybridized Triboelectric Nanogenerator. *ACS Nano* **2014**, *8*, 1932–1939. [CrossRef] [PubMed]
50. Zhang, H.; Yang, Y.; Su, Y.; Chen, J.; Adams, K.; Lee, S.; Hu, C.; Wang, Z.L. Triboelectric Nanogenerator for Harvesting Vibration Energy in Full Space and as Self-Powered Acceleration Sensor. *Adv. Funct. Mater.* **2014**, *24*, 1401–1407. [CrossRef]
51. Wang, X.; Niu, S.; Yin, Y.; Yi, F.; You, Z.; Wang, Z.L. Triboelectric Nanogenerator Based on Fully Enclosed Rolling Spherical Structure for Harvesting Low-Frequency Water Wave Energy. *Adv. Energy Mater.* **2015**, *5*, 1501467. [CrossRef]
52. Seol, M.L.; Han, J.W.; Jeon, S.B.; Meyyappan, M.; Choi, Y.K. Floating Oscillator-Embedded Triboelectric Generator for Versatile Mechanical Energy Harvesting. *Sci. Rep.* **2015**, *5*, 16409. [CrossRef] [PubMed]
53. Ahmed, A.; Saadatnia, Z.; Hassan, I.; Zi, Y.; Xi, Y.; He, X.; Zu, J.; Wang, Z.L. Self-Powered Wireless Sensor Node Enabled by a Duck-Shaped Triboelectric Nanogenerator for Harvesting Water Wave Energy. *Adv. Energy Mater.* **2017**, *7*. [CrossRef]
54. Su, Y.; Wen, X.; Zhu, G.; Yang, J.; Chen, J.; Bai, P.; Wu, Z.; Jiang, Y.; Wang, Z.L. Hybrid triboelectric nanogenerator for harvesting water wave energy and as a self-powered distress signal emitter. *Nano Energy* **2014**, *9*, 186–195. [CrossRef]
55. Lin, Z.-H.; Cheng, G.; Wu, W.; Pradel, K.C.; Wang, Z.L. Dual-Mode Triboelectric Nanogenerator for Harvesting Water Energy and as a Self-Powered Ethanol Nanosensor. *ACS Nano* **2014**, *8*, 6440–6448. [CrossRef] [PubMed]
56. Shi, Q.; Wang, H.; Wu, H.; Lee, C. Self-powered triboelectric nanogenerator buoy ball for applications ranging from environment monitoring to water wave energy farm. *Nano Energy* **2017**, *40*, 203–213. [CrossRef]
57. Cui, N.; Gu, L.; Lei, Y.; Liu, J.; Qin, Y.; Ma, X.; Hao, Y.; Wang, Z.L. Dynamic Behavior of the Triboelectric Charges and Structural Optimization of the Friction Layer for a Triboelectric Nanogenerator. *ACS Nano* **2016**, *10*, 6131–6138. [CrossRef] [PubMed]
58. Gooding, D.M.; Kaufman, G.K. Tribocharging and the Triboelectric Series. *Encycl. Inorg. Bioinorg. Chem.* **2014**, 1–9. [CrossRef]
59. Lee, J.-W.; Hwang, W. Theoretical study of micro/nano roughness effect on water-solid triboelectrification with experimental approach. *Nano Energy* **2018**, *52*, 315–322. [CrossRef]
60. Niu, S.; Liu, Y.; Wang, S.; Lin, L.; Zhou, Y.S.; Hu, Y.; Wang, Z.L. Theoretical Investigation and Structural Optimization of Single-Electrode Triboelectric Nanogenerators. *Adv. Funct. Mater.* **2014**, *24*, 3332–3340. [CrossRef]
61. Niu, S.; Liu, Y.; Chen, X.; Wang, S.; Zhou, Y.S.; Lin, L.; Xie, Y.; Wang, Z.L. Theory of freestanding triboelectric-layer-based nanogenerators. *Nano Energy* **2015**, *12*, 760–774. [CrossRef]

© 2018 by the authors. Licensee MDPI, Basel, Switzerland. This article is an open access article distributed under the terms and conditions of the Creative Commons Attribution (CC BY) license (http://creativecommons.org/licenses/by/4.0/).

Article

Manipulation of *p-/n*-Type Thermoelectric Thin Films through a Layer-by-Layer Assembled Carbonaceous Multilayer Structure

Wonjun Jang [1], Hyun A Cho [1], Kyungwho Choi [2,*] and Yong Tae Park [1,*]

1. Department of Mechanical Engineering, Myongji University, Yongin, Gyeonggi 17058, Korea; wjang@mju.ac.kr (W.J.); chocho9508@hanmail.net (H.A.C.)
2. New Transportation Innovative Research Center, Korea Railroad Research Institute, 176, Cheoldobangmulgwan-ro, Uiwang-si, Gyeonggi-do 16105, Korea
* Correspondence: kwchoi80@krri.re.kr (K.C.); ytpark@mju.ac.kr (Y.T.P.); Tel.: +82-31-460-5603 (K.C.); +82-31-330-6343 (Y.T.P.)

Received: 30 October 2018; Accepted: 25 November 2018; Published: 28 November 2018

Abstract: Recently, with the miniaturization of electronic devices, problems with regard to the size and capacity of batteries have arisen. Energy harvesting is receiving significant attention to solve these problems. In particular, the thermoelectric generator (TEG) is being studied for its ability to harvest waste heat energy. However, studies on organic TEGs conducted thus far have mostly used conductive polymers, making the application range of TEGs relatively narrow. In this study, we fabricated organic TEGs using carbonaceous nanomaterials (i.e., graphene nanoplatelet (GNP) and single-walled carbon nanotube (SWNT)) with polyelectrolytes (i.e., poly(vinyl alcohol) (PVA) and poly (diallyldimethyl ammonium chloride) (PDDA)) via layer-by-layer (LbL) coating on polymeric substrates. The thermoelectric performance of the carbonaceous multilayer structure was measured, and it was confirmed that the thermoelectric performance of the TEG in this study was not significantly different from that of the existing organic TEG fabricated using the conductive polymers. The 10 bilayer SWNT thin films with polyelectrolyte exhibited a thermopower of $-14~\mu V \cdot K^{-1}$ and a power factor of $25~\mu W \cdot m^{-1} K^{-2}$. Moreover, by simply changing the electrolyte, *p*- or *n*-type TEGs could be easily fabricated with carbonaceous nanomaterials via the LbL process. Also, by just changing the electrolyte, *p*- or *n*-type of TEGs could be easily fabricated with carbonaceous nanomaterials with a layer-by-layer process.

Keywords: thermoelectric; layer-by-layer; graphene; carbon nanotube; thin film

1. Introduction

The internet of things (IoT) and wearable devices market is experiencing rapid growth. As a result, the demand for batteries to be used in these devices is increasing; however, increasing miniaturization of electronic devices is limited by the replacement of batteries and explosion risk [1,2]. To solve these problems, various approaches have been studied, among which energy harvesting is currently the most significant.

Energy can be harvested by triboelectric [3], thermoelectric [4], or piezoelectric [5] techniques, and extensive thermoelectric research is underway owing to the advantages of harvesting waste heat. The thermoelectric generator (TEG) harvests electricity via the Seebeck effect, which states that thermoelectric power is generated when there is a temperature gradient in the system. Traditional TEGs have been fabricated using inorganic semiconductors such as bismuth telluride to maximize the Seebeck effect [6–8]. However, the conventional thermoelectric materials possess the disadvantages of being hard, deformable, toxic, and expensive. Therefore, studies regarding organic thermoelectric materials have been actively conducted to overcome these limitations [9,10]. An organic TEG is basically composed of conductive polymers such as polyaniline (PANi) or poly(3,4-ethylenedioxythiophene) doped with poly(styrene sulfonate) anions (PEDOT:PSS). It has a low thermal conductivity and can be deposited on a flexible substrate, indicating that conductive materials are advantageous thermoelectric materials. Moreover, the incorporation of carbon-based nanofillers in the polymer matrix can enhance or manipulate the electrical properties of the nanocomposites. Therefore, studies have been focused on enhancing the thermoelectric effect using composites with carbonaceous nanomaterials such as graphene or carbon nanotubes (CNTs) [11–16]. Several methods exist to fabricate polymeric composites with carbonaceous nanomaterials. In this study, polymer-CNT and polymer-graphene multilayers were fabricated using a simple, inexpensive, and versatile method.

The layer-by-layer (LbL) technique, which has been studied for a long time as one of the wet coating methods, has gained attention as a very simple and low-cost process [17,18]. In particular, it is possible to fabricate flexible thin films of various materials on a substrate, such as cotton, which is difficult to coat using conventional methods. Using the LbL method, various properties such as flame retardancy [19], electrical conductivity [20], or sensor characteristics [21], can be imparted to the desired materials. However, only a few papers have observed the thermoelectric properties of the LbL-assembled thin film TEGs [22–25]. In this study, the thermoelectric properties of graphene or CNT thin films coated on a poly(ethylene terephthalate) (PET) substrate were analyzed.

Carbonaceous nanomaterials with a high electrical conductivity, such as graphene and CNTs, have been studied with organic/inorganic composite TEGs combined with polymers. Since these studies aim to increase the thermoelectric performance of an organic TEG, conductive polymers (e.g., PANi, PEDOT:PSS, etc.) were mostly used for organic materials. In this study, the thermoelectric properties of the TEG were investigated using poly(vinyl alcohol) (PVA) and poly(diallyldimethyl ammonium chloride) (PDDA) without a conducting polymer. In addition, using the LbL technique, it was possible to readily fabricate a carbonaceous nanomaterial-polymer composite TEG, enabling the wider use of organic TEGs.

2. Materials and Methods Methods

2.1. Materials

Purified electric arc single-walled carbon nanotubes (SWNTs, TUBALL™ SWNT, individual tube: average 1 µm length and 2 nm diameter, carbon content 75%) and graphene nanoplatelets (GNPs, Angstron, maximum X–Y dimensions of 10 µm, carbon content 95%, oxygen content ≤ 2.5%) were used in this study. PDDA (M_w~200,000–350,000 g/mol, Figure 1a), sodium deoxycholate (DOC, $C_{24}H_{39}NaO_4$, Figure 1b), PVA (M_w~89,000–98,000 g/mol, Figure 1c), poly(4-styrenesulfonic acid) (PSS, M_w~75,000 g/mol, 18 wt% in H_2O, Figure 1d), isopropyl alcohol (IPA), and methanol were purchased from Sigma-Aldrich (Yongin, Korea). All chemicals were used as received. A PET film (100 µm thickness, Goodfellow, Huntingdon, England, UK) and single-side-polished silicon wafers (University Wafer, Boston, MA, USA) were purchased as substrates.

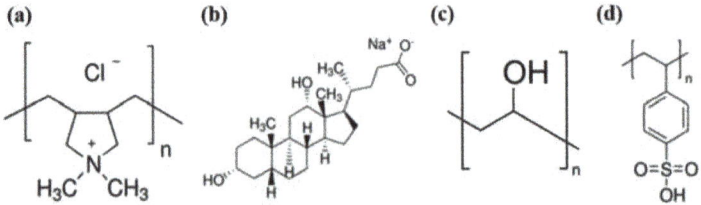

Figure 1. Chemical structures of (**a**) poly(diallyldimethyl ammonium chloride) (PDDA), (**b**) sodium deoxycholate (DOC), (**c**) poly(vinyl alcohol) (PVA), and (**d**) poly(4-styrenesulfonic acid) (PSS).

2.2. Layer-by-Layer Assembly

The LbL technique was used in this study to uniformly coat GNPs or SWNTs on the PET substrate (see Figure 2). First, 0.05 wt% SWNTs were added to deionized (DI) water with 1.5 wt% DOC and dispersed for 60 min using a tip sonicator (UW2070, Banderin Electronic, Berlin, Germany). CNTs require special chemicals to disperse in water owing to their entanglement and hydrophobicity. DOC is widely known as a surfactant that aids in the proper dispersion of SWNTs in deionised (DI) water [26]. The prepared SWNT-DOC solution had a negative charge [27]. Subsequently, 0.25 wt% PDDA was added to DI water and dispersed for 30 min using a bath sonicator (Branson, CPX-3800H, Emerson Electric Company, Ferguson, MO, USA). The PDDA solution had a positive charge, which serves as a counterpart of the SWNT-DOC solution. The PET film, which was surface-cleaned using a plasma etcher (Harrick Plasma, PDC 32G-2, Ithaca, NY, USA) for 5 min, was initially applied to PDDA, a positively charged solution, for 5 min. After rinsing and drying, it was placed in the SWNT-DOC solution, which was a negatively charged solution. After 5 min, rinsing and drying were repeated, by which the PDDA and SWNT-DOC layers were uniformly deposited on the PET substrate by charge bonding [20,27]. One bilayer (BL) was generated by the first cycle and consisted of one layer of PDDA and SWNT-DOC. Thus, in the case of n BL coating, PDDA and SWNT-DOC multilayers were coated on the PET substrate, and this was labeled as PET[PDDA/SWNT-DOC]$_n$. Similarly, the GNP multilayers were coated on a PET substrate as follows: 0.1 wt% GNP was added to DI water with 0.1 wt% PSS and dispersed for 180 min using a tip sonicator. PSS is widely known as a dispersing agent to help disperse GNP properly in DI water. This is because the sulfonic functional groups of PSS prevent the aggregation of GNP in DI water [28]. As a counterpart of the GNP-PSS solution, 0.25 wt% PVA was added to DI water and dispersed via magnetic stirring at 70 °C for 30 min. PVA forms a hydrogen bond with PSS, allowing the stable formation of multilayers of GNP-PSS and PVA [28,29]. Similarly, each layer of PVA and GNP-PSS was labeled as 1 BL and a PVA and GNP-PSS multilayer film coated on the PET substrate was labeled as PET[PVA/GNP-PSS]$_n$.

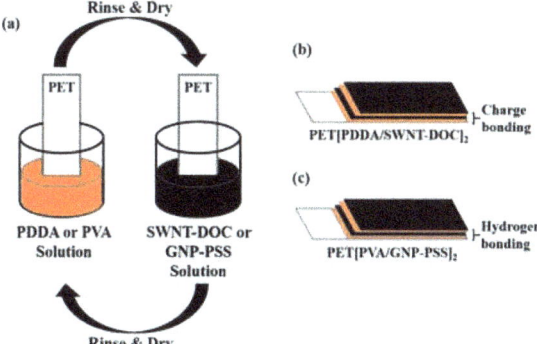

Figure 2. (a) Schematics of layer-by-layer process for PET[PDDA/SWNT-DOC] and PET[PVA/GNP-PSS] thin films. (b) Schematics of a PET[PDDA/SWNT-DOC]$_2$ sample based on charge bonding. (c) Schematics of a PET[PVA/GNP-PSS]$_2$ sample based on hydrogen bonding.

2.3. Characterization

LbL thin films on a PET substrate were used to analyze UV–Visible (UV–Vis, Ocean optics, Largo, FL, USA) light absorbance. The thicknesses of the LbL films deposited on the thermally oxidized silicon wafers were determined using a spectroscopic ellipsometer (SE, V-VASE, J.A. Woollam Co., Lincoln, NE, USA). Surface images of the LbL-coated PET samples were observed using a field emission scanning electron microscope (FE-SEM) device (SU-70, Hitachi, Tokyo, Japan) at 15 kV. Images for cross-sections of both samples were obtained using a transmission electron microscope (TEM, JEM-2100F, JEOL, Tokyo, Japan) to characterize the carbonaceous nanomaterial-polymer multilayers on PET.

For thermoelectric performance measurement, a PET[PVA/GNP-PSS] or PET[PDDA/SWNT-DOC] LbL sample (width: 10 mm, length: 55–60 mm) was installed between two Peltier devices (Marlow industries, Dallas, TX, USA) acting in opposite directions. The Peltier device initiates thermoelectric measurement by providing a temperature gradient to the test sample. The temperature gradient of the sample can be measured with a thermocouple composed of copper and constantan. The Peltier device operates in the temperature range of −4 to +4 K, and the maximum temperature difference range is approximately 7 K. Subsequently, to measure the electrical resistance and voltage of the specimen, a silver paste was applied to the surface of the specimen to minimize the electrical contact resistance. Thereafter, with a four-point probe, the electrical resistance was measured using a pair of electrodes, and the potential difference was measured by connecting a voltmeter to the other pair. The electrical resistance was obtained from the I-V curve, and the thermopower was obtained from the V-T curve.

3. Results

Figure 3a shows the thickness of the [PDDA/SWNT-DOC] and [PVA/GNP-PSS] LbL thin films. The thickness increases with an increase in the number of BLs in both systems. SWNT LbL thin films, with a thickness of approximately 7 nm at 3 BLs, increase in thickness to 20 nm at 10 BLs. This suggests that SWNT is linearly coated as the number of BLs increases. The GNP LbL thin film possesses a thickness of 30 nm at 3 BLs, which increases to 82 nm at 10 BLs. It also demonstrates that GNPs are linearly coated with increasing BLs. The SWNT LbL film is thinner than the GNP sample, because GNP is a 2-D platelet and SWNT is a 1-D tube [27,29]. As the number of BLs increases, the GNP LbL thin film becomes relatively thicker than the SWNT film, as shown in the SEM and TEM images in Figure 3b,c.

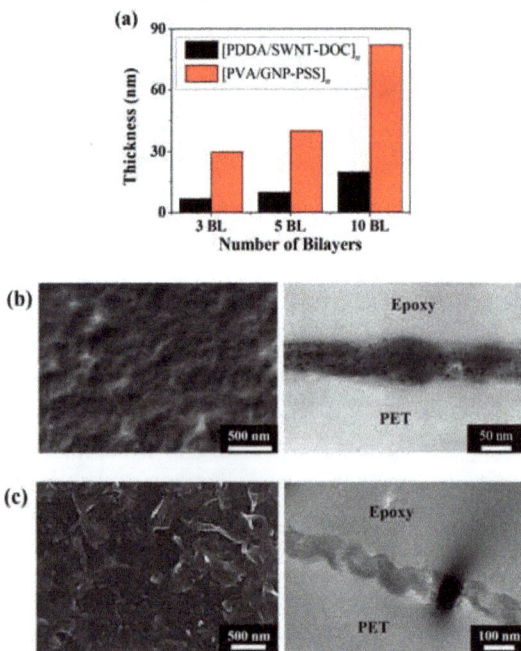

Figure 3. (a) Ellipsometric thickness of [PDDA/SWNT-DOC]$_n$ and [PVA/GNP-PSS]$_n$ (n = 3, 5, and 10) thin films. Both increase uniformly as the number of BLs increases. SEM surface images and TEM cross-sectional images of (b) [PDDA/SWNT-DOC]$_{20}$ and (c) [PVA/GNP-PSS]$_{20}$ thin films.

Figure 4a,b show the light absorbance of PET[PDDA/SWNT-DOC] and PET[PVA/GNP-PSS] with an increasing number of BLs. As shown in Figure 4c, the light absorbance increases uniformly from 1 to 10 BLs in both thin films. This indicates that SWNT and GNP layers are linearly deposited during LbL coating. In addition, it can be confirmed that SWNT bilayers exhibit a relatively much lower absorbance than GNP because SWNT is a 1-D material with an exceptionally small diameter and GNP is a relatively larger 2-D platelet. Based on its characteristics, SWNT is widely used as a transparent electrode in several studies [30–32]. Figure 4c,d show the absorbance and transmittance at 550 nm in the two LbL systems. As can be observed from Figure 4c, the absorbance linearly increases to 0.16 for PET[PDA/SWNT-DOC]$_{10}$ and to 0.9 for PET[PVA/GNP-PSS]$_{10}$. Conversely, the light transmittance decreases to 69% for PET[PDA/SWNT-DOC]$_{10}$ and 11.6% for PET[PVA/GNP-PSS]$_{10}$. This can be observed more clearly in Figure 4e, which shows 3, 5, and 10 BL PET[PDA/SWNT-DOC] and PET[PVA/GNP-PSS] samples with a bare PET (0 BL), respectively. Both carbonaceous nanomaterials (i.e., SWNT and GNP) demonstrate that as the number of BLs increases, the color of the LbL thin films gradually darkens; in particular, the GNP LbL thin film is distinctly darker.

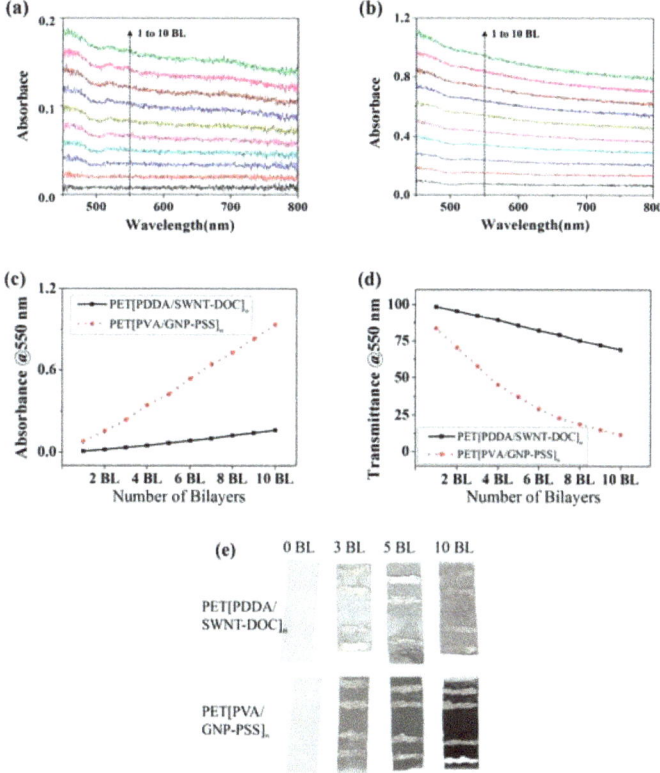

Figure 4. (a) Visible light absorbance spectra of PET[PDDA/SWNT-DOC]$_n$ (n = 1–10) LbL thin films. (b) Visible light absorbance spectra of PET[PVA/GNP-PSS]$_n$ (n = 1–10) LbL thin films. (c) Visible light absorbance of PET[PDDA/SWNT-DOC]$_n$ (n = 1–10) (solid line) and PET[PVA/GNP-PSS]$_n$ (n = 1–10) (dotted line) at 550 nm wavelength. (d) Visible light transmittance of PET[PDDA/SWNT-DOC]$_n$ (n = 1–10) (solid line) and PET[PVA/GNP-PSS]$_n$ (n = 1–10) (dotted line) at 550 nm wavelength. (e) Images of PET[PDDA/SWNT-DOC]$_n$ (n = 3, 5, 10) and PET[PVA/GNP-PSS]$_n$ (n = 3, 5, 10) with control (0 BL) samples. In the LbL samples above, the silver material on each sample surface is silver paste applied for four-point probe measurement.

Figure 5a shows the electrical resistance of PET[PDDA/SWNT-DOC] LbL thin films at 3, 5, and 10 BLs. As a result of the ellipsometric thickness and light absorbance, the amount of SWNT coated on the PET increases as the number of BLs increases, so that the absorbance and thickness also increase. Likewise, the electrical resistance decreases as the number of BLs deposited increases. The electrical resistance of PET[PDDA/SWNT-DOC] thin films decreases from 1.3 kΩ to 0.31 kΩ at 3 to 10 BLs, respectively. Similarly, Figure 5b also shows the decrease in the electrical resistance of PET[PVA/GNP-PSS] thin films according to the increase in the number of BLs deposited. The electrical resistance decreases from 29.5 kΩ to 5.0 kΩ at 3 to 10 BLs, respectively. In this case, it can be confirmed that the electrical resistance of the SWNT LbL thin film is much lower than that of the GNP LbL thin film, because the SWNT can form a 3-D network with an increasing number of deposited BLs and can enable considerably efficient electron transport. This can be more distinctly observed from the previous SEM and TEM images. Further, the GNP LbL thin film exhibits a relatively higher electrical resistance because it is stacked in a 2-D shape [3,29,33]. Figure 5c,d show the electrical conductivity obtained by applying geometrical factors to the measured resistance values. GNP LbL thin films with a relatively increased thickness and a 2-D network structure adversely affect the carrier transport, demonstrating a conductivity of approximately 10 S·cm^{-1}, but the SWNT samples demonstrated an increase of 2 orders. In both samples, the conductivity increased with increasing number of BLs owing to the formation of more pathways for electrical carriers. Figure 5e,f show the thermopower values of SWNT and GNP LbL thin films, respectively. For SWNT and GNP, the values of thermopower are negative and positive, respectively. This is because the PDDA/SWNT-DOC multilayers are n-type thermoelectric materials with electrons being the major carrier and PVA/GNP-PSS multilayers are p-type thermoelectric materials. SWNT LbL thin films have a constant thermopower value of approximately -14 μV·K^{-1}, whereas that of GNP is approximately 15 μV·K^{-1}. This is an interesting result because a majority of the n-type organic nanocomposites demonstrate a relatively low conductivity compared to p-type composites due to the conversion of the major carrier from holes to electrons. Free-standing organic nanocomposites with PEI-doped SWNT exhibit 1–10 S·cm^{-1} [34] and LbL multilayers of a PEI-doped double-walled carbon nanotubes composite demonstrate up to 300 S·cm^{-1} of conductivity with 80 BLs [35]. In this research, an electrical conductivity of 10^3 S·cm^{-1} was achieved through only 3 to 10 BLs of SWNTs with a negative thermopower of ~15 μV·K^{-1}. Although the doping mechanism and carrier transportation process is beyond this research, it is valuable to report this result since a high current output in the thermoelectric generation system is a problem in industrial applications of organic thermoelectric materials.

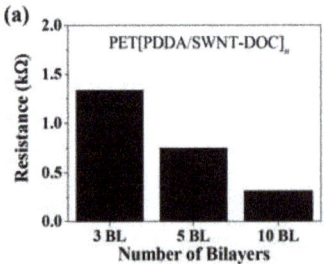
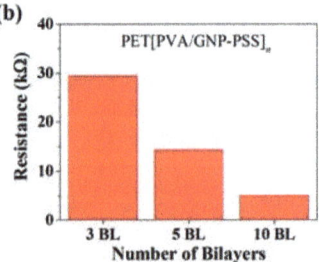

Figure 5. *Cont.*

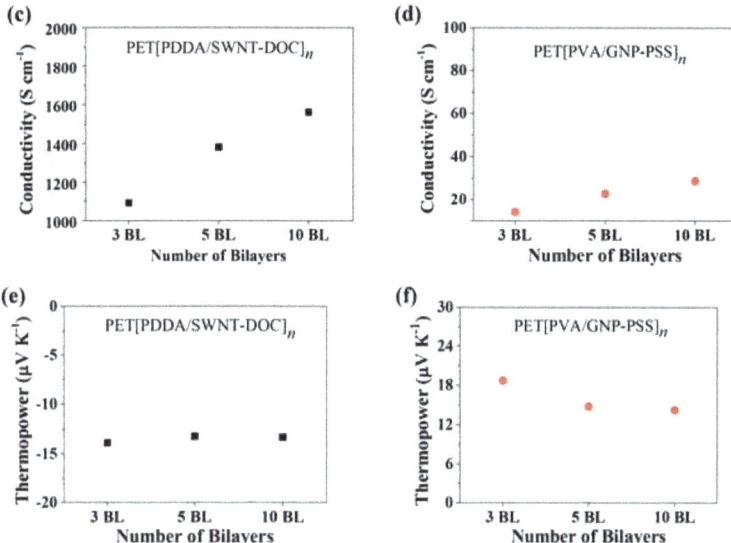

Figure 5. (**a,b**) Electrical resistance of PET[PDDA/SWNT-DOC]$_n$ (n = 3, 5, 10) and PET[PVA/GNP-PSS]$_n$ (n = 3, 5, 10) LbL thin films. (**c,d**) Electrical conductivity of PET[PDDA/SWNT-DOC]$_n$ (n = 3, 5, 10) and PET[PVA/GNP-PSS]$_n$ (n = 3, 5, 10) LbL thin films. (**e,f**) Thermopower (Seebeck coefficient) of PET[PDDA/SWNT-DOC]$_n$ (n = 3, 5, 10) and PET[PVA/GNP-PSS]$_n$ (n = 3, 5, 10) LbL thin films.

In Figure 6, with the measured electrical conductivity and thermopower, the power factor (P.F. = $S^2 \cdot \sigma$) was calculated as a function of the number of BLs. Although the absolute values of the thermopower of both SWNT and GNP LbL thin films are similar, the power factor of the 10 BL SWNT sample was 28 $\mu W \cdot m^{-1} K^2$, whereas that of the GNP sample was less than 1 $\mu W \cdot m^{-1} K^2$. This is comparable to other carbonaceous organic nanocomposites with an exceedingly high conductivity, even among p-type organic thermoelectric materials [36–38].

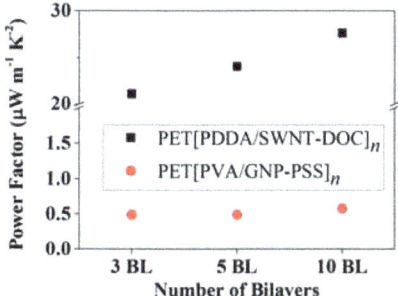

Figure 6. Thermoelectric power factor of PET[PDDA/SWNT-DOC]$_n$ (n = 3, 5, 10) and PET[PVA/GNP-PSS]$_n$ (n = 3, 5, 10) LbL thin films.

4. Conclusions

Carbonaceous nanomaterial-polymer multilayer composites (i.e., [PDDA/SWNT-DOC]$_n$ and [PVA/GNP-PSS]$_n$) were applied to fabricate cost-effective, scalable, and flexible TEGs, where high performance and high versatility (p- and n-type TEGs) were demonstrated. Alternate exposure of the PET substrate to two opposite aqueous solutions yielded LbL thin films with a linear trend of growth, evidenced by the thickness and light absorbance. [PDDA/(SWNT-DOC)]$_{10}$ films possess a conductivity, thermopower, and power factor of 1.8×10^3 S·cm^{-1}, -14 μV·K^{-1} (n-type thermoelectric material), and 28 μW·m^{-1}K^2, respectively. Compared with previous organic/inorganic composite TEGs fabricated using conventional conductive polymers, the thermopower of the LbL thin films in this study was very similar. The SWNT-based thin film was a flexible TEG, and is capable of replacing conventional brittle TEGs in certain energy harvesting applications. Moreover, the carbonaceous nanomaterial-polymer composite TEG is expected to lower the cost of TEG fabrication and enable fabrication using a wider variety of materials. Further study of other types of LbL materials (i.e., polymers, dispersions, etc.) would further reduce the electrical resistance and increase the thermopower of thin-film TEGs.

Author Contributions: W.J., K.C., and Y.T.P. conceptualized the phenomenon. Y.T.P. and K.C. designed the project. W.J., H.A.C., and K.C. conducted the experiments. All authors wrote the manuscript and analyzed the data. Y.T.P. and K.C. supervised the whole project.

Funding: This work was financially supported by the Mid-career Researcher Program (No. 2017R1A2B4006104) through a National Research Foundation of Korea (NRF) grant funded by the Ministry of Science, ICT and Future Planning (MSIP). This work was also supported by the Korea Institute of Energy Technology Evaluation and Planning (KETEP) and the Ministry of Trade, Industry & Energy (MOTIE) of the Republic of Korea (No. 20174010201160).

Conflicts of Interest: The authors declare no conflict of interest.

References

1. Shastry, S.; Langdorf, M.I. Electronic vapor cigarette battery explosion causing shotgun-like superficial wounds and contusion. *West J. Emerg. Med.* **2016**, *17*, 177. [CrossRef] [PubMed]
2. Masih-Tehrani, M.; Ha'iri-Yazdi, M.R.; Esfahanian, V.; Safaei, A. Optimum sizing and optimum energy management of a hybrid energy storage system for lithium battery life improvement. *J. Power Sources* **2013**, *244*, 2–10. [CrossRef]
3. Chung, I.J.; Kim, W.; Jang, W.; Park, H.W.; Sohn, A.; Chung, K.B.; Kim, D.W.; Choi, D.; Park, Y.T. Layer-by-layer assembled graphene multilayers on multidimensional surfaces for highly durable, scalable, and wearable triboelectric nanogenerators. *J. Mater. Chem. A* **2018**, *6*, 3108–3115. [CrossRef]
4. Kim, D.; Kim, Y.; Choi, K.; Grunlan, J.C.; Yu, C. Improved thermoelectric behavior of nanotube-filled polymer composites with poly (3, 4-ethylenedioxythiophene) poly (styrenesulfonate). *ACS Nano* **2009**, *4*, 513–523. [CrossRef] [PubMed]
5. Yang, Y.; Zhang, H.; Zhu, G.; Lee, S.; Lin, Z.H.; Wang, Z.L. Flexible hybrid energy cell for simultaneously harvesting thermal, mechanical, and solar energies. *ACS Nano* **2012**, *7*, 785–790. [CrossRef] [PubMed]
6. Venkatasubramanian, R.; Silvola, E.; Colpitts, T.; O'Quinn, B. Thin-film thermoelectric devices with high room-temperature figures of merit. In *Materials for Sustainable Energy: A Collection of Peer-Reviewed Research and Review Articles from Nature Publishing Group*; World Scientific: London, UK, 2011; pp. 120–125.
7. Poudel, B.; Hao, Q.; Ma, Y.; Lan, Y.; Minnich, A.; Yu, B.; Yan, X.; Wang, D.; Muto, A.; Vashaee, D.; et al. High-thermoelectric performance of nanostructured bismuth antimony telluride bulk alloys. *Science* **2008**, *320*, 634–638. [CrossRef] [PubMed]
8. Heremans, J.P.; Jovovic, V.; Toberer, E.S.; Saramat, A.; Kurosaki, K.; Charoenphakdee, A.; Yamanaka, S.; Snyder, G.J. Enhancement of thermoelectric efficiency in PbTe by distortion of the electronic density of states. *Science* **2008**, *321*, 554–557. [CrossRef] [PubMed]
9. Sun, Y.; Sheng, P.; Di, C.; Jiao, F.; Xu, W.; Qiu, D.; Zhu, D. Organic thermoelectric materials and devices based on p- and n-type poly (metal 1,1,2,2-ethenetetrathiolate)s. *Adv. Mater.* **2012**, *24*, 932–937. [CrossRef] [PubMed]
10. Zhang, Q.; Sun, Y.; Xu, W.; Zhu, D. Organic thermoelectric materials: Emerging green energy materials converting heat to electricity directly and efficiently. *Adv. Mater.* **2014**, *26*, 6829–6851. [CrossRef] [PubMed]
11. Du, Y.; Shen, S.Z.; Yang, W.; Donelson, R.; Cai, K.; Casey, P.S. Simultaneous increase in conductivity and Seebeck coefficient in a polyaniline/graphene nanosheets thermoelectric nanocomposite. *Synth. Met.* **2012**, *161*, 2688–2692. [CrossRef]

12. Yan, H.; Kou, K. Enhanced thermoelectric properties in polyaniline composites with polyaniline-coated carbon nanotubes. *J. Mater. Sci.* **2014**, *49*, 1222–1228. [CrossRef]
13. Ma, P.C.; Tang, B.Z.; Kim, J.K. Conversion of semiconducting behavior of carbon nanotubes using ball milling. *Chem. Phys. Lett.* **2008**, *458*, 166–169. [CrossRef]
14. Mo, C.; Jian, J.; Lia, J.; Fanga, Z.; Zhaoa, Z.; Yuana, Z.; Yanga, M.; Zhanga, Y.; Dai, L.; Yu, D. Boosting water oxidation on metal-free carbon nanotubes via directional interfacial charge-transfer induced by an adsorbed polyelectrolyte. *Energy Environ. Sci.* **2018**. [CrossRef]
15. Duan, Y.; Holmes, N.E.; Ellard, A.L.; Gao, J.; Xue, W. Solution-based fabrication and characterization of a logic gate inverter using random carbon nanotube networks. *IEEE Trans. Nanotechnol.* **2013**, *12*, 1111–1117. [CrossRef]
16. Blackburn, J.L.; Ferguson, A.J.; Cho, C.; Grunlan, J.C. Carbon-Nanotube-Based Thermoelectric Materials and Devices. *Adv. Mater.* **2018**, *30*, 1704386. [CrossRef] [PubMed]
17. Decher, G.; Lvov, Y.; Schmitt, J. Proof of multilayer structural organization in self-assembled polycation-polyanion molecular films. *Thin Solid Films* **1994**, *244*, 772–777. [CrossRef]
18. Hammond, P.T. Form and function in multilayer assembly: New applications at the nanoscale. *Adv. Mater.* **2004**, *16*, 1271–1293. [CrossRef]
19. Choi, K.; Seo, S.; Kwon, H.; Kim, D.; Park, Y.T. Fire protection behavior of layer-by-layer assembled starch-clay multilayers on cotton fabric. *J. Mater. Sci.* **2018**, *53*, 11433–11443. [CrossRef]
20. Moon, G.; Jang, W.; Son, I.; Cho, H.A.; Park, Y.T.; Lee, J.H. Fabrication of new liquid crystal device using layer-by-layer thin film process. *Processes* **2018**, *6*, 108. [CrossRef]
21. Park, J.J.; Hyun, W.J.; Mun, S.C.; Park, Y.T.; Park, O.O. Highly stretchable and wearable graphene strain sensors with controllable sensitivity for human motion monitoring. *ACS Appl. Mater. Interfaces* **2015**, *7*, 6317–6324. [CrossRef] [PubMed]
22. Culebras, M.; Cho, C.; Krecker, M.; Smith, R.; Song, Y.; Gomez, C.M.; Cantarero, A.; Grunlan, J.C. High thermoelectric power factor organic thin films through combination of nanotube multilayer assembly and electrochemical polymerization. *ACS Appl. Mater. Interfaces* **2017**, *9*, 6306–6313. [CrossRef] [PubMed]
23. Cho, C.; Bittner, N.; Choi, W.; Hsu, J.H.; Yu, C.; Grunlan, J.C. Thermally enhanced n-type thermoelectric behavior in completely organic graphene oxide-based thin films. *Adv. Electron. Mater.* **2018**, 1800465. [CrossRef]
24. Cho, C.; Wallace, K.L.; Tzeng, P.; Hsu, J.H.; Yu, C.; Grunlan, J.C. Outstanding low temperature thermoelectric power factor from completely organic thin films enabled by multidimensional conjugated nanomaterials. *Adv. Energy Mater.* **2016**, *6*, 1502168. [CrossRef]
25. Cho, C.; Stevens, B.; Hsu, J.H.; Bureau, R.; Hagen, D.A.; Regev, O.; Yu, C.; Grunlan, J.C. Completely organic multilayer thin film with thermoelectric power factor rivaling inorganic tellurides. *Adv. Mater.* **2015**, *27*, 2996–3001. [CrossRef] [PubMed]
26. Moriarty, G.P.; Wheeler, J.N.; Yu, C.; Grunlan, J.C. Increasing the thermoelectric power factor of polymer composites using a semiconducting stabilizer for carbon nanotubes. *Carbon* **2012**, *50*, 885–895. [CrossRef]
27. Park, Y.T.; Ham, A.Y.; Grunlan, J.C. High electrical conductivity and transparency in deoxycholate-stabilized carbon nanotube thin films. *J. Phys. Chem. C* **2010**, *114*, 6325–6333. [CrossRef]
28. Lee, S.W.; Park, J.J.; Park, B.H.; Mun, S.C.; Park, Y.T.; Liao, K.; Seo, T.S.; Hyun, W.J.; Park, O.O. Enhanced sensitivity of patterned graphene strain sensors used for monitoring subtle human body motions. *ACS Appl. Mater. Interfaces* **2017**, *7*, 11176–11183. [CrossRef] [PubMed]
29. Mun, S.C.; Park, J.J.; Park, Y.T.; Kim, D.Y.; Lee, S.W.; Cobos, M.; Ye, S.J.; Macosko, C.W.; Park, O.O. High electrical conductivity and oxygen barrier property of polymer-stabilized graphene thin films. *Carbon* **2017**, *125*, 492–499. [CrossRef]
30. Tenent, R.C.; Barnes, T.M.; Bergeson, J.D.; Ferguson, A.J.; To, B.; Gedvilas, L.M.; Heben, M.J.; Blackburn, J.L. Ultrasmooth, large-area, high-uniformity, conductive transparent single-walled-carbon-nanotube films for photovoltaics produced by ultrasonic spraying. *Adv. Mater.* **2009**, *21*, 3210–3216. [CrossRef]
31. Takenobu, T.; Takahashi, T.; Kanbara, T.; Tsukagoshi, K.; Aoyagi, Y.; Iwasa, Y. High-performance transparent flexible transistors using carbon nanotube films. *Appl. Phys. Lett.* **2006**, *88*, 033511. [CrossRef]
32. Shim, B.S.; Zhu, J.; Jan, E.; Critchley, K.; Kotov, N.A. Transparent conductors from layer-by-layer assembled SWNT films: Importance of mechanical properties and a new figure of merit. *ACS Nano* **2010**, *4*, 3725–3734. [CrossRef] [PubMed]

33. Park, Y.T.; Ham, A.Y.; Grunlan, J.C. Heating and acid doping thin film carbon nanotube assemblies for high transparency and low sheet resistance. *J. Mater. Chem.* **2011**, *21*, 363–368. [CrossRef]
34. Freeman, D.D.; Choi, K.; Yu, C. N-type thermoelectric performance of functionalized carbon nanotube-filled polymer composites. *PLoS ONE* **2012**, *7*, e47822. [CrossRef] [PubMed]
35. Cho, C.; Culebras, M.; Wallace, K.L.; Song, Y.; Holder, K.; Hsu, J.H.; Yu, C.; Grunlan, J.C. Stable n-type thermoelectric multilayer thin films with high power factor from carbonaceous nanofillers. *Nano Energy* **2016**, *28*, 426–432. [CrossRef]
36. Yu, C.; Kim, Y.S.; Kim, D.; Grunlan, J.C. Thermoelectric behavior of segregated-network polymer nanocomposites. *Nano Lett.* **2008**, *8*, 4428–4432. [CrossRef] [PubMed]
37. Lee, W.; Hong, C.T.; Kwon, O.H.; Yoo, Y.; Kang, Y.H.; Lee, J.Y.; Cho, S.Y.; Jang, K.S. Enhanced thermoelectric performance of bar-coated SWCNT/P3HT thin films. *ACS Appl. Mater. Interfaces* **2015**, *7*, 6550–6556. [CrossRef] [PubMed]
38. Wang, L.; Yao, Q.; Bi, H.; Huang, F.; Wang, Q.; Chen, L. PANI/graphene nanocomposite films with high thermoelectric properties by enhanced molecular ordering. *J. Mater. Chem. A* **2015**, *3*, 7086–7092. [CrossRef]

© 2018 by the authors. Licensee MDPI, Basel, Switzerland. This article is an open access article distributed under the terms and conditions of the Creative Commons Attribution (CC BY) license (http://creativecommons.org/licenses/by/4.0/).

Review

Recent Progress in Flexible Organic Thermoelectrics

Mario Culebras [1], Kyungwho Choi [2] and Chungyeon Cho [3],*

1. Stokes Laboratories, Bernal Institute, University of Limerick, Limerick, Ireland; mario.culebrasrubio@ul.ie
2. Transportation Innovative Research Center, Korea Railroad Research Institute, Uiwang-si 16105, Korea; kwchoi80@krri.re.kr
3. Department of Carbon Convergence Engineering, College of Engineering, Wonkwang University, Iksan 54538, Korea
* Correspondence: cncho37@wku.ac.kr; Tel.: +82-63-850-7274

Received: 29 October 2018; Accepted: 25 November 2018; Published: 30 November 2018

Abstract: Environmental energy issues caused by the burning of fossil fuel such as coal, and petroleum, and the limited resources along with the increasing world population pose a world-wide challenge. Alternative energy sources including solar energy, wind energy, and biomass energy, have been suggested as practical and affordable solutions to future energy needs. Among energy conversion technologies, thermoelectric (TE) materials are considered one of the most potential candidates to play a crucial role in addressing today's global energy issues. TE materials can convert waste heat such as the sun, automotive exhaust, and industrial processes to a useful electrical voltage with no moving parts, no hazardous working chemical-fluids, low maintenance costs, and high reliability. These advantages of TE conversion provide solutions to solve the energy crisis. Here, we provide a comprehensive review of the recent progress on organic TE materials, focused on polymers and their corresponding organic composites incorporated with carbon nanofillers (including graphene and carbon nanotubes). Various strategies to enhance the TE properties, such as electrical conductivity and the Seebeck coefficient, in polymers and polymer composites will be highlighted. Then, a discussion on polymer composite based TE devices is summarized. Finally, brief conclusions and outlooks for future research efforts are presented.

Keywords: thermoelectric; graphene; carbon nanotubes; power factor; polymers; energy harvesting; organic composites

---

1. Introduction

The global demand for sustainable and renewable energy sources due to climate change caused by burning fossil fuels has incurred interest in developing new types of energy [1]. Various renewable resources including solar energy, wind, geothermal, and wave have been investigated to produce power without giving rise to any carbon dioxide emissions, but their performance has still been limited [2,3]. Considering that more than 60% of global energy is produced by cars, transportation, and mechanical systems is wasted as the form of heat, reconverting this wasted heat into useful electricity can be a viable way in addressing current global energy issues [4–6]. The direct conversion between waste heat and electrical power based on thermoelectric (TE) materials has attracted tremendous attention because they have various applications such as eco-friendly power generation and cooling systems with no moving parts and pollution-free step.

The efficiency of TE materials is evaluated by a dimensionless figure-of-merit, ZT, which is defined as $S^2 \cdot \sigma \cdot T \cdot \kappa^{-1}$, where S, σ, κ, and T are the Seebeck coefficient, electrical conductivity, thermal conductivity, and absolute temperature of the material, respectively. The power factor (PF), defined as $S^2 \cdot \sigma$, is occasionally used as an alternative to ZT because of the low κ for polymer-based TE materials or difficulties in the precise in-plane κ measurement for thin film thickness (<1 µm) [7,8].

A high-performance TE material should possess a high S to promote the energy conversion of heat to electricity, a high σ to minimize Joule heating, and a low κ to keep a temperature gradient [9]. However, it is very difficult to modulate independently these three parameters because of a strong interdependence between them in traditional bulk materials, which imposes a limitation on maximizing ZT as a whole. For example, increasing σ is usually accompanied by a decreased S and an increased carrier contribution to κ.

Recent advances in engineering the electronic structure of TE nanostructures have led to a significant improvement of ZT of >1 by preferentially reducing the lattice κ by involving the phonon scattering effect within superlattices or nanoinclusion [10]. Moreover, the energy filtering effect, quantum confinement, and tuning the electronic band structure (i.e., the density of the states) through nanostructures engineering have been shown to independently improve S without greatly suppressing σ (hence, the enhancement of PF) [11–13]. When the system size approaches a scale comparable to the feature length of electron behavior (i.e., mean free path, wavelength) in any direction, low-dimensional materials could create the spike-like shape of electronic density of states (e.g., Van Hove singularities), which could result in increased asymmetry of the differential conductivity at around the Fermi energy $\left[\frac{\partial DOS(E)}{\partial E}\right]_{E=EF}$ [11].

Currently, the most widely used TE materials are conventional inorganic semiconductors based on metal alloys (i.e., Bi_2Te_3, SiGe, and PbTe) due to their high PF [14]. Besides their high TE efficiency, the inorganic-based TE materials have been used to power space vehicles because of their suitability for high-temperature (600–1000 °C) power generation [15]. Despite their promise, they possess certain inherent limitations such as a scarcity of materials, high cost of production, brittleness, the difficulty of processing, and toxicity that hinder the full utilization of their unique benefits. Organic materials could alleviate these problems because of their advantageous properties, including having a light weight, low cost, mechanical flexibility, convenience to be processed, and solution processability over large areas [16,17]. Moreover, compared to semiconductor-based TE materials, organic materials possess an intrinsically low κ, typically ranging from 0.1 to 1 $W \cdot m^{-1} \cdot K^{-1}$, which is beneficial to the enhancement of TE performance [8,18]. Although there has been progress, the PF of conducting polymers is low relative to inorganic materials, and their ZT is 2–3 orders of magnitude lower than that of the inorganic TE materials because of their low σ and S, which hampers their use in TE applications [19]. Recently, organic TE nanocomposites have been exhibited to achieve high TE performance by judicious combination with low-dimensional carbon nanofillers such as graphene or carbon nanotubes (CNTs). Incorporating carbon materials into polymer matrices results in synergies, such as a high σ and S, while retaining low κ [20].

In this review, the recent advances in the TE performances of polymers and their corresponding organic composites comprised of carbon nanofillers, including graphene, graphene oxide, and carbon nanotubes, are highlighted. We discuss the preparation strategies to optimize the TE properties and organic TE devices are also summarized. Finally, a brief conclusion has been provided along with a perspective and outlook on the future development of flexible organic TE materials.

2. Polymer-Based Thermoelectric Materials

Since the discovery of iodine-doped polyacetylene that conducts like a metal by MacDiarmid [21], intrinsically conducting polymers have been widely explored for various applications in the area of optoelectronic devices, supercapacitors, biosensors, photovoltaic cells, batteries and electromagnetic shielding [22,23]. This is related to the set of advantageous properties associated with their unique structures with excellent physical and chemical properties [24]. Among the various conducting polymers, considerable attention has been recently paid to polyaniline (PANi) and poly(3,4-ethylenedioxythiophene)-poly(styrenesulfonate) (PEDOT:PSS), Poly(3-hexylthiophene) (P3HT), and their derivatives as promising candidates of TE materials. The chemical structure of intrinsically conductive polymers, along with the insulating polymers discussed in this review, is shown in Table 1. A summary of the TE properties of polymers and their composites compounded with carbon nanofillers is provided in Table 2. In this section, we have highlighted the TE performance of conducting polymers such as PANi, PEDOT and its derivatives, and P3HT.

Table 1. The chemical structure of polymers used to prepare thermoelectric materials.

Nonconducting Polymers	Abbreviation	Polymer Structure	Conducting Polymers	Abbreviation	Polymer Structure
Polypropylene	PP		polypyrrole	PPy	
polyvinylpyrrolidone	PVP		polythiophene	PTh	
poly(vinyl acetate)	PVAc		poly(3,4-ethylenedioxy thiophene)	PEDOT	
polyethylenimine	PEI		Polyaniline leucoemeraldine (x=1, y=0) emeraldine(x=y=0.5) pernigraniline(x=0, y=1)	PANI	
Nafion	Nafion		Poly(3,4-ethylenedioxy thiophene)-poly(styrenesulfonate)	PEDOT:PSS	

Table 2. The thermoelectric properties of polymers and carbon nanofiller-filled polymer composites.

Materials	Systems	Electrical Conductivity (S·cm^{-1})	Seebeck Coefficient (μV·K^{-1})	Power Factor (μW·m^{-1}·K^{-2})	Thermal Conductivity (W·m^{-1}·K^{-1})	ZT	Ref.
Non-Conducting Polymers	PVP	5 × 10^{-4}	−9.5	5 × 10^{-6}			[25]
	PEI	5.5 × 10^{-4}	−5.5	2 × 10^{-6}			[25]
Conducting polymers	PANi	173	14	3.5			[26]
		67	12.8	1.1			[27]
		270	20	10.8			[28]
		600			0.276	2.7 × 10^{-4} (423 K)	[29]
	PEDOT:PSS	800	65	400	0.52	0.42 (300 K)	[30]
		1900	20.6	80.63	0.2	0.32	[31]
				318.4	0.3	0.31	[32]
		2980	21.9	142			[33]
	P3HT	0.47	386	7			[34]
		320	269	62.4			[35]
	PPy	4.6	9.4	0.04			[16]
All-Polymer Composites	PEDOT/PEDOT:Tos			446.6			[31]
	PEDOT:PSS/P3HT	200	17	5.79		0.44 (300 K)	[36]
	PEDOT:PSS/PANi-CSA			56			[37]
Graphene-based Composites	PANi/Graphene	1.2 × 10^4	18.66	4.2	1.2	1.37 × 10^{-3} (393 K)	[38]
		856	5.6	30.9			[39]
		814	19	55			[40]
		750	26	58.8		4.84 × 10^{-4}	[41]
			28				[42]
	PEDOT:PSS/Graphene	637	26.8	11		2.1 × 10^{-2}	[43]
				624			[44]
				45.7			[45]
				83.2		0.1	[46]
	P3HT/Graphene	1.2	35.5	0.16			[47]
	PPy/Graphene	75	34	10.2	0.84	2.8 × 10^{-3}	[48]
				8.6			[49]
	Ppy/Graphene/PANi	500	32	52.5			[50]

Table 2. Cont.

Materials	Systems	Electrical Conductivity (S·cm^{-1})	Seebeck Coefficient (μV·K^{-1})	Power Factor (μW·m^{-1}·K^{-2})	Thermal Conductivity (W·m^{-1}·K^{-1})	ZT	Ref.
	Nafion/CNTs	13	28	1			[51]
	PVAc/CNTs	48			0.34	0.006	[52]
	PVAc/SWNT-GA	90.4	40	14.5	0.25		[53]
	PVAc/MWNT-TCPP	1.28	28	0.1			[54]
	PVAc/DWNT-TCPP	71	78	42.8			[55]
		61	28.6	5			[56]
		1.6	26	1.1			[57]
	PANi/CNT	16	10	0.16		7.6 × 10^{-5}	[58]
		31.5	45.4	6			[59]
		125	40	20	1.5	0.004	[60]
		1440	39	217			[61]
CNT-based Composites		400	25	25	0.4	0.02 (300 K)	[62]
	PEDOT:PSS/CNT	1350	41	160			[63]
		960	70	500			[64]
		241	38.9	21.1			[65]
		780	43.7	150			[66]
	P3HT/MWNT	30	28	2.4	0.59	8.7 × 10^{-4}	[67]
				105			[68]
	P3HT/SWNT	1000	32.5	107			[69]
		2760	31	308			[70]
	P3HT/CNT	348	97	325			[71]
	PPy/MWNT	72	17	2			[72]
	PPy/SWNT			21.7			[73]
Graphene/CNT-based Composites	PEDOT:PSS/Graphene/MWNT	690	23	37			[74]
	PEDT/rGO/SWNT	208	21	9			[75]
	PANi/Graphene/PANi/DWNT	1050	132	1825	0.36	0.31 (300 K)	[76]
	PANi/Graphene-PEDOT: PSS/PANi/DWNTPEDOT:PSS	1900	120	2710			[77]
n-type TE Materials	poly[Kx (Ni-ett)]	15	−100	66			[78]
	PVAc/PEI/CNT-SDBS	240	−80	15		0.2	[79]
	PEI/CNT-NaBH$_4$	300	−80	153.6			[80]
	DWNT-PEI/Graphene-PVP			190			[81]
	PP/SWNT/CuO/PEG	0.024	−57	0.008			[82]
	BV/CNT	2220	−116	3100			[83]

2.1. Polyaniline (PANi)

Polyaniline (PANi) has been considered one of the most promising conductive polymers with merits of the structure diversification, environmental stability, and unique doping/dedoping process. The very first investigation on the TE properties of PANi was done by Yan and Toshima in 1999 [26]. They fabricated multilayered thin films by alternately layering electrically insulating emeraldine base layers and electrically conducting (±)-10-camphorsulfonic acid (CSA)-doped emeraldine salt layers. By doping the films with CSA, multilayers consisting of the two kinds of PANi exhibited an electrical conductivity of 173 S·cm^{-1} and Seebeck coefficient of 14 µV·K^{-1}, translating to the power factor of 3.5 µW·m^{-1}·K^{-2} at 300 K. This value was 3.5 times higher relative to that of the bulk counterpart. The same research group also investigated the anisotropy of the TE properties in CSA-doped PANi films [27]. In their study, the PANi molecules took an ordered and extended coil-like conformation upon the stretching process, resulting in enhancement in carrier mobility and hence, electrical conductivity and Seebeck coefficient in the PANi films. The stretched CSA-doped PANi films exhibited higher values in ZT (2.9×10^{-2}) compared to that of the unstretched counterpart (4.4×10^{-3}).

The TE properties of PANi can be modulated by controlling its structure [84]. The submicron-fiber structure of PANi was created by doping with CSA and followed by mixing with m-cresol. PANi(CSA)/m-cresol had a great quantity of submicron fibers with a typical size of 200–500 nm in diameter and 5–10 µm in length. PANi(CSA) samples with no m-cresol treatment displayed a general grain structure. PANi(CSA)/m-cresol exhibited the electrical conductivity of 67 S·cm^{-1} and the Seebeck coefficient of 12.8 µV·K^{-1}. This resulted in a power factor of 1.1 µW·m^{-1}·K^{-2} at 300 K, which was about 20 times higher than that of PANi(CSA). The higher TE properties in PANi(CSA)/m-cresol over PANi(CSA) were due to structural changes of PANi that CSA-doped PANi formed liquid-crystalline solutions in m-cresol with an extraordinary order of chain packing [85], which was confirmed with X-ray diffraction (XRD) patterns. Therefore, the submicron-fiber structure where the oriented PANi chains were clustered had less π-conjugation defects in the polymer backbone, which was beneficial for improving the carrier mobility and, hence, the electrical conductivity.

Li et al. studied the effects of hydrochloric acid (HCl)-doping concentrations on the TE properties of PANi [29]. HCl-doped PANi films prepared by chemical oxidative polymerization exhibited that elevated HCl-doping levels led to an increase in electrical conductivity up to 600 S·cm^{-1} at 1.25 M and then a dramatic reduction above this concentration. The Seebeck coefficient showed an opposite trend to that of the electrical conductivity in which a maximum value was obtained at 0.25 M. Due to its mainly amorphous structure, the thermal conductivity of PANi was in the range of 0.1 W·m^{-1}·K^{-1} [86]. In this work, the thermal conductivity was insensitive to doping conditions, showing 0.276 W·m^{-1}·K^{-1} at 303 K. The maximum ZT reached 2.7×10^{-4} at 423 K when the HCl-doping concentration was 1.0 M.

Modulating the ordered degree of polymeric molecular chains is an effective way to realize the synergistic regulation of electrical conductivity and Seebeck coefficient and therefore improve the TE properties of conducting polymers [28]. Yao et al. investigated the intrinsic effect of the molecular structure on the electric transport of PANi by tuning the molecular chain alignment. The strong van der Waals attractions between polymer chains in most solvents caused PANi to adopt a compacted coil conformation. However, the chemical interactions between CSA-doped PANi and m-cresol induced the molecular structure of PANi to change to an expanded coil, which was confirmed with the ultraviolet-visible (UV-vis) spectra in Figure 1a. This structural change of PANi to the expanded molecular conformation reduced the π defects caused by ring twisting and strengthened the π-π conjugation interactions between rings and therefore enhanced the delocalization of carriers along the polymer chain. The ordered chain packing increased the carrier mobility and improved both the electrical conductivity and Seebeck coefficient (and hence, the power factor) with the m-cresol content, as shown in Figure 1b,c.

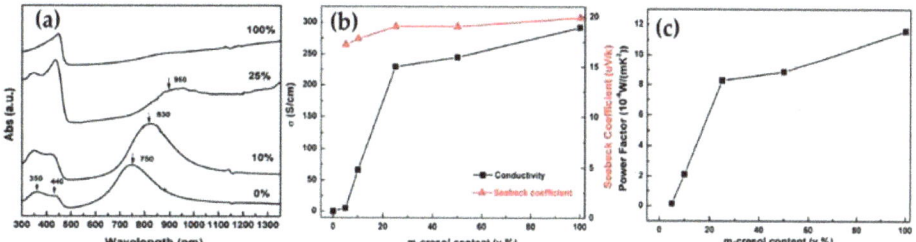

Figure 1. (**a**) The ultraviolet-visible (UV-vis) spectra of polyaniline (PANi) dissolved in different m-cresol contents, (**b**) the electrical conductivity and Seebeck coefficient, and (**c**) the power factor of the polyaniline films prepared with different m-cresol contents. Reprinted with permission from [28]. Copyright 2014 The Royal Society of Chemistry.

2.2. Poly(3,4-ethylenedioxythiophene) (PEDOT) and Their Derivatives

The electrically conductive polymer PEDOT, one of the polythiophene derivatives, has gained particular attention due to their highly conductive properties, optical transparency, solution processability, and remarkable capabilities of easy doping [87]. The highly electro-conducting PEDOT films have been used as effective organic materials for many applications including organic solar cells, light emitting diodes, and organic field-effect transistors. The TE properties of PEDOT have been shown to be controlled precisely via chemical or electrochemical doping that enhances electrical conductivity without significantly compromising the Seebeck coefficient, thus achieving the improved TE properties. For instance, Bubnova et al. demonstrated the optimization of ZT in PEDOT:p-toluenesulfonate (PEDOT:Tos) by using tetrakis(dimethylamino)ethylene (TDAE) as the de-doping agent [88]. Upon exposure of a PEDOT:Tos film into a vapor of TDAE molecules, electron transfer from the reducing agent to the polymer transformed oxidized PEDOT:Tos chains into neutral ones. At the optimum oxidation level of 22%, they reported a very high power factor of 324 $\mu W \cdot m^{-1} \cdot K^{-2}$ and a low thermal conductivity of 0.37 $W \cdot m^{-1} \cdot K^{-2}$, yielding a ZT = 0.25 at room temperature. Another high TE performance of PEDOT films was prepared through the precise control of the oxidation level of the polymer electrochemically [89]. A mixture of pyridine and PEG-b-poly(propylene glycol)-b-PEG tri-block copolymer was used as mediators for the polymerization of 3,4-ethylenedioxythiophene (EDOT) in the presence of Fe-tosylate (Tos). A maximum power factor of 1270 $\mu W \cdot m^{-1} \cdot K^{-2}$ was obtained when the applied potential was 0.1 V.

Of the various PEDOTs available, doping with PSS is the most studied PEDOT derivative because of its inherent properties such as water-solubility, low cost, flexible mechanical properties, thermal stability, high transparency, and commercial availability, making it suitable for TE applications. Although the electrical conductivity of PEDOT:PSS has been shown to be enhanced by a variety of organic solvents such as dimethyl sulfoxide (DMSO), N,N-dimethylformamide (DMF), and tetrahydrofuran (THF), the reported ZT values were as low as 10^{-4} to 10^{-2} [90,91]. However, Kim et al. reported a maximum ZT value of 0.42 by mixing PEDOT:PSS with 5% of DMSO and ethylene glycol (EG) in 2013 [30]. The incorporation of 0.2 wt% of PEDOT nanowires into PEDOT-based polymer hosts by in situ polymerization led to a power factor as high as 446.6 $\mu W \cdot m^{-1} \cdot K^{-2}$ and the ZT value was up to 0.44 at room temperature [92].

Thermoelectric performance of PEDOT:PSS can be enhanced by treatment with secondary dopants. Free-standing flexible and smooth PEDOT:PSS buckypapers, prepared using vacuum-assisted filtration, revealed that the electrical conductivity was enhanced to 1900 $S \cdot cm^{-1}$ by treating PEDOT:PSS with formic acid [31]. Since the secondary dopants did not change the oxidation level of PEDOT [93], the Seebeck coefficient remained relatively constant. The power factor of the formic acid-treated films was up to 80.6 $\mu W \cdot m^{-1} \cdot K^{-2}$ and the thermal conductivity, measured with the Harman method, was 0.2 $W \cdot m^{-1} \cdot K^{-1}$, which translated to a ZT value as high as 0.32 at room temperature. The enhancement of TE properties in the PEDOT:PSS films was mainly due to the selective removal of PSS chains by secondary dopants with high dielectric constants. The coulombic interaction between the positively charged PEDOT and negatively charged PSS chains were reduced through screening effect of formic acid, which in turn facilitated the removal of non-conductive PSS and led to a three-dimensional conjugated network of highly conjugated PEDOT [94].

The TE properties of PEDOT can be finely controlled by doping it with various counter-ions. Culebras et al. investigated the TE performances of different PEDOT derivatives including PEDOT:ClO$_4$, PEDOT:PF$_6$, and PEDOT:bis(tifluoromethylsulfonyl)imide (BTFMSI), which were synthesized via electro-polymerization [95]. The electrical conductivity was sensitive to the type of dopants, while the Seebeck coefficient remained at the same order of magnitude (Figure 2a–c). The larger size of the counter-ion was more favorable for PEDOT polymers to change from the typical coil conformation to linear or expanded-coil polymeric structures due to the electrostatic interaction between the positive charges of PEDOT and negative charges of counter-ions, as shown in Figure 2d. Therefore, by increasing the size of the counter-ion, the electrical conductivity increased in the order of PEDOT:BTFMSI, PEDOT:PF$_6$, and PEDOT:ClO$_4$. The TE properties of PEDOT derivatives were further optimized upon the chemical reduction with hydrazine. By reducing them with hydrazine for only 5 s, PEDOT:BTFMSI films exhibited a power factor of 147 $\mu W \cdot m^{-1} \cdot K^{-2}$ and a thermal conductivity of 0.19 $W \cdot m^{-1} \cdot K^{-1}$, which translated to a ZT of 0.22 at room temperature.

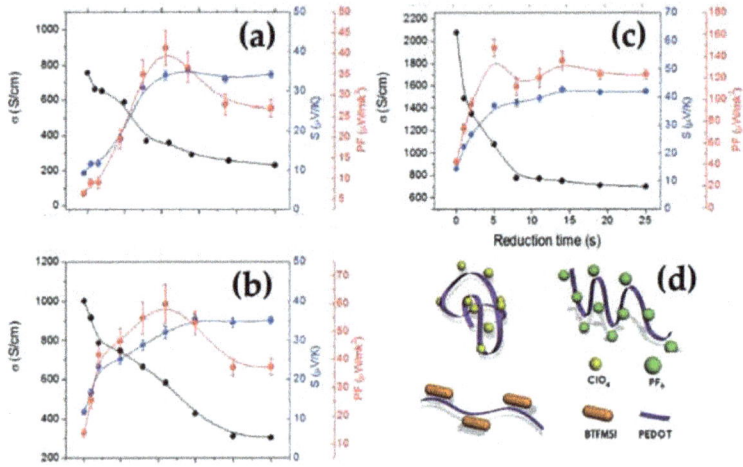

Figure 2. The thermoelectric properties of poly(3,4-ethylenedioxythiophene) (PEDOT):ClO$_4$ (a), PEDOT:PF$_6$ (b), and PEDOT: bis(tifluoromethylsulfonyl)imide (BTFMSI) (c) as a function of chemical reduction time using hydrazine. PEDOT conformation in the presence of different counter-ions (d). Reprinted with permission from [95]. Copyright 2014 The Royal Society of Chemistry.

Wang et al. reported the effect of the shear printing parameters on the electric transport mechanism in PEDOT films via the natural brush-printing method [96]. They investigated the interplay between backbone alignment, aggregation, and charge transport anisotropy in semiconducting polymers. It was found that the shear-induced charge transport is closely related to the conformational changes of polymer backbones, aggregation, and crystallization during the film formation process. The imposed shear stress enhanced polymer chain alignment, thereby facilitating aggregation and improving charge transport.

Optimizing the oxidation level of PEDOT:PSS films via sequential doping and dedoping can be a promising route toward high TE systems. Lee et al. reported that the synergistic effect of sequential doping with TSA and followed by dedoping with hydrazine/DMSO is an effective way to improve the TE properties by precisely controlling the PSS concentration and the oxidation level of PEDOT. In their study, the treatment of the chemical dopant p-toluenesulfonic acid monohydrate (TSA) on PEDOT:PSS films drove the positively charged holes to more effectively move through the polymer and to facilitate the formation of polarons, consequently increasing the carrier concentration, as shown in Figure 3a [32]. The Coulombic attraction between PEDOT and PSS was weakened because protonated hydrazine ions ($N_2H_5^+$) were associated with negatively charged PSS (Figure 3b). Furthermore, the dedoping process with the hydrazine/DMSO treatment during spin coating selectively removed insulating PSS molecules in the films. Although the hydrazine/DMSO dedoping decreased the electrical conductivity due to a reduction in the carrier concentration, the Seebeck coefficient dramatically increased. As a result, a power factor was up to 318.4 $\mu W \cdot m^{-1} \cdot K^{-2}$ and the thermal conductivity decreased from 0.38 to 0.30 upon the removal of PSS, which translated to a ZT value of 0.31 at room temperature.

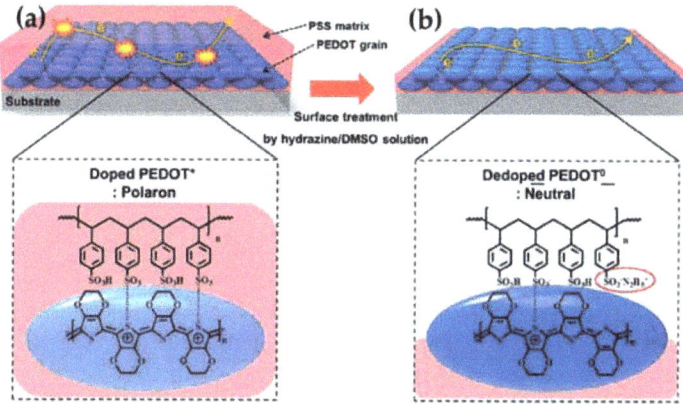

Figure 3. An overall scheme for the spin-coated PEDOT:poly(styrenesulfonate) (PSS) film fabricated by the sequential treatment of p-toluenesulfonic acid monohydrate (TSA)/dimethyl sulfoxide (DMSO) doping and hydrazine/DMSO dedoping. (**a**) DMSO/TSA-doped PEDOT:PSS film formation via spin coating and selective removal of PSS and the dedoped PEDOT:PSS (DDTP) film treated with hydrazine/DMSO solution (**b**). Reprinted with permission from [32]. Copyright 2014 The Royal Society of Chemistry.

A post-treatment with various reagents has been shown to enhance the TE performance. Zhu et al. found out that the TE properties of PEDOT:PSS films were enhanced by employing a deep eutectic solvent (DES), which is a new generation of green solvents having merits such as biocompatibility, non-toxicity, biodegradability, and chemical inertness with water [97,98]. After dropping DES, a mixture of choline chloride and ethylene glycol in the ratio of 1:3, onto PEDOT:PSS prepared by the casting method, the PEDOT:PSS films exhibited an improvement in electrical conductivity to 620 $S \cdot cm^{-1}$ and in the Seebeck coefficient to 29.1 $\mu V \cdot K^{-1}$. At an optimized condition (at 100 °C for 10 h heating) of DES treatment, the power factor reached 24 $\mu W \cdot m^{-1} \cdot K^{-2}$, which was approximately four orders of magnitude higher relative to the pure PEDOT:PSS. Atomic force microscope (AFM) and X-ray photoelectron spectroscopy (XPS) measurements revealed that the remarkably enhanced electrical conductivity originated from the removal of the excess insulating PSS and the phase separation between the PEDOT and PSS chains. The removal of the excess insulating PSS resulted in a continuous conducting network of the PEDOT-rich phase, which in turn improved the electrical conductivity.

A superacid, trifluoromethanesulfonic acid (TFMS), can lead to enhance the TE properties of the PEDOT:PSS films. Wang et al. demonstrated that the TFMS treatment in methanol (MeOH) on PEDOT:PSS films induced the removal of the insulating PSS and polymer chain rearrangements, giving, in turn, a denser packing of the conductive PEDOT polymer chains [33]. During the dedoping process, TFMS-MeOH as a post-treatment induced the chemical reaction between CF_3SO_3H and PSS to generate non-nucleophilic $CF_3SO_3^-$ anion and the PSSH polyacid, followed by removal of PSSH and the formation of a densely packed crystalline structure. TEMS-MeOH post-treatment increased the electrical conductivity to 2980 S·cm^{-1} and the Seebeck coefficient to 21.9 µV· K^{-1}, translating to a power factor of 142 µW·m^{-1}·K^{-2}.

2.3. Poly(3-hexylthiophene) (P3HT)

Poly(3-hexylthiophene) (P3HT) has been widely used in the fabrication of optoelectronic devices due to its excellent electrical properties, appropriate energy gap, and doping reversibility. Easy processability in the solution state makes P3HT suitable for a variety of solution processes such as spray-printing, spin-coating, roll-to-roll printing, and inkjet printing. Zhu et al. investigated the TE performances of P3HT films doped with iodine vapor that were prepared by casting the P3HT solution [34]. Iodine-doped P3HT was conducted by an electron exchange process between iodine as an electron acceptor and P3HT as an electron donor. Upon iodine vapor treatment on pristine P3HT films, the polymer chains self-organized into a more ordered structure, which was confirmed with AFM analysis. They demonstrated that the size of P3HT aggregates was reduced and the number of P3HT single chains increased after doping with iodine vapor, which induced P3HT chains to self-organize into a more ordered structure [34]. As a result, electron transport was enhanced, improving the TE performance. The maximum electrical conductivity of iodine-doped P3HT films was 4.7×10^{-1} S·cm^{-1}, which was five orders of magnitude higher relative to that of the counterpart. A comparatively high Seebeck coefficient of 386 µV· K^{-1} was obtained. The calculated power factor was estimated to be 7 µW·m^{-1}·K^{-2} at room temperature, which was higher than that of the counterpart.

The charge carriers in the conducting polymers transport through the inter-chain and intra-chain hopping processes. Regulating the configuration and arrangement of polymer chains directly affects the carrier transport and, consequently, TE performances. Qu et al. prepared anisotropic P3HT films with a highly oriented morphology using 1,3,5-trichlorobenzene (TCB), an organic small-molecules as a template for polymer epitaxy [35]. In general, P3HT chains adopt an irregular configuration and random arrangement largely due to the flexible hexyl side chains [99], which decreases the carrier mobility and deteriorates the TE performances. However, the resulting P3HT films via a temperature-gradient crystallization process using an organic small-molecule, TCB, as the template for polymer epitaxy exhibited a reduction in π-π conjugation defects along the polymer backbone and also effectively increased the degree of electron delocalization. This produced a large-scale and quasi-1D pathway for the carrier movement and resulted in an enhanced carrier mobility. The electrical conductivity and Seebeck coefficient of the TCB-treated P3HT films were up to 320 S·cm^{-1} and 269 µV· K^{-1}, respectively. As a result, the power factor and ZT value at 365 K reached 62.4 µW·m^{-1}·K^{-2} and 0.1, respectively, in the direction parallel to the fiber axis.

2.4. Polymer/Polymer Thermoelectric Composites

Compounding two or more polymers into one system can be an effective approach for enhancing the TE properties by taking advantages of each polymer. For example, Zhang et al. investigated TE properties of PEDOT nanowire/PEDOT hybrids where PEDOT nanowires synthesized by template-confined in situ polymerization were incorporated into PEDOT:PSS and PDOT:tosylate (Tos), respectively [30]. The TE properties of the hybrids exhibited that the power factor was enhanced by 9 times relative to PEDOT:PSS mixed with 5 vol% DMSO while the low thermal conductivity was maintained. The large enhancement of the TE performance resulted from the synergistic effect of interfacial energy filtering, the nanowire percolation threshold, and the possible changes of carrier concentration in the PEDOT:PSS host materials. Upon the addition of 0.2 wt% PEDOT nanowires to PEDOT:Tos nanocomposites, the power factor of the hybrid materials was increased to 446.6 µW·m^{-1}·K^{-2} and the ZT reached 0.44 at room temperature.

A novel generation of bilayered nanofilms in which a pure organic PEDOT:PSS nanofilm was obtained via spin-coating techniques and then, followed by depositing P3HT using electrochemical polymerization displayed a good electrochemical stability and enhanced TE performance [36]. Early theoretical simulations indicated that the TE properties of the hybrid composites could not exceed the maximum of each component, but possessed an intermediate between them [37,100]. However, this work by Shi et al. showed that the TE performance could be higher relative to the parent films it was composed of, which could be due to the energy filtering effect created by a potential barrier in nanostructured structures. The electrical conductivity of the resultant PEDOT:PSS/P3HT nanofilms reached up to 200 S·cm^{-1} along with the Seebeck coefficient of 17 µV·K^{-1} at 300 K, which translated to a power factor of 5.79 µW·m^{-1}·K^{-2}.

The multilayer structures composed of PEDOT:PSS and PANi-CSA polymers assembled by a layer-by-layer deposition using a spin-coating method enhanced the TE power factor [101]. The multilayer films displayed an enhanced electrical conductivity without sacrificing the Seebeck coefficient, which is typically true in traditional bulk inorganic materials. A PEDOT:PSS/PANi-CSA multilayer exhibited a synergistic improvement in the electrical conductivity, which was likely ascribed to stretching of the backbone chains in both the PEDOT:PSS and PANi-CSA layers. Hole diffusion due to a difference in the Fermi energy between polymers could enhance electrical conductivity. The 4 layers of PEDOT:PSS/PANi-CSA films exhibited a power factor of 56 µW·m^{-1}·K^{-2}.

3. Carbon-Based Polymer Nanocomposites

Although the conducting polymers have been shown to be very attractive thermoelectric (TE) materials due to easy processing and environmentally-benign characteristics, along with low thermal conductivity, which is ideal for TE efficiency, their TE properties are still inferior to those of conventional inorganics, primarily due to the low electrical conductivity and Seebeck coefficient [102]. Recently, incorporating carbon nanofillers such as carbon nanotubes (CNTs) and graphene, into the polymer matrix has been proved to be an effective strategy to enhance the electrical conductivity and Seebeck coefficient, resulting in a significant improvement in power factor [17,103]. The unique TE behaviors (i.e., decoupled TE physical parameters) and synergistic enhancement effects have been found in polymer/carbon composites owing to the combination of the advantages of each component and the electrical/thermal transport behaviors at the numerous interfaces [7,102]. In this section, we focus on TE properties of polymers/carbon nanofillers, including graphene and CNTs composites.

3.1. Graphene-Based Nanocomposites

3.1.1. PANi-Based Graphene Composites

Wang et al. investigated the TE properties of the HClO$_4$-doped PANi/graphite composites, prepared by ball milling and cold pressing, as a function of graphite concentration [38]. Although the thermal conductivity increased with the graphite content, both the electrical conductivity and Seebeck coefficient showed a significant improvement, leading to a large enhancement in the TE performances. When the graphite content increased from 0 to 50 wt%, the electrical conductivity exhibited a dramatic improvement from 1.23×10^2 to 1.2×10^4 S·cm^{-1} and the Seebeck coefficient also increased from 0.82 to 18.66 µV·K^{-1}. The decoupled TE behavior was attributed to the numerous interfaces between the HClO$_4$-doped PANi and graphite, which was formed during ball milling. With increasing graphite, there was an obvious improvement in power factor from 8.3×10^{-5} to 4.2 µW·m^{-1}·K^{-2}. Therefore, the ZT of the PANi/graphite composites containing a graphite concentration of 50 wt% reached 1.37×10^{-3} at 393 K, which was four orders of magnitude higher than the HClO$_4$-doped PANi.

The TE bulk composites pellets and films in which PANi and graphene nanosheets (GNs) were mixed with various ratios displayed an interesting TE behavior [39]. When the weight ratio of PANi to GNs decreased from 4:1 to 1:1, both the electrical conductivity and the Seebeck coefficient simultaneously increased in both pellets and films, which could be explained by a dramatic increase in carrier mobility, while the carrier concentration remained unchanged. The power factor of the pellets was 5.6 µW·m^{-1}·K^{-2}. PANi/GNs nanocomposites, prepared by in situ polymerization of aniline monomer in the presence of GNs, were studied by Xiang et al [104]. They fabricated a paper-like nanocomposite in which PANi grew on the basal plane of GNs via a strong π interaction using controlled vacuum filtration of an aqueous dispersion of PANi. Although GNs are easily agglomerated in the aqueous solution due to the layer structure and the large specific surface area of graphene, the strong π-π interaction between the nuclei of PANi and GNs helped overcome the van der Waals attraction and facilitate uniform dispersion of GNs. GNs served as the template, helped PANi chains more ordered and aligned on the GNs. The aligned PANi chain conformation reduced the carrier hopping resistance, consequently leading to an improvement in carrier mobility and hence Seebeck coefficient. The chemically stretched PANi nanofibril on the surface of GNs as a result of π electron interaction enhanced TE properties, producing a ZT value of 1.5×10^{-4} at room temperature.

The dispersion state, the structural defects, and the impurity element content in the graphene can influence the TE properties of the PANi/graphene nanocomposites [40]. PANi/graphene composites, prepared by a solution-assisted dispersing method, showed higher TE properties with graphene having lower structural defects and oxygen impurities. The maximum electrical conductivity and power factor of the composite reached 856 S·cm^{-1} and 19 μW·m^{-1}·K^{-2}. These outstanding TE properties of PANi/graphene composites were attributed to the strong π-π conjugation interactions between PANi and low-defect graphene. The strong electronic interaction between PANi and graphene induced PANi chains to orient along the surface of graphene with more ordered conformation, which minimized the π-π conjugation defects in PANi molecular chains and resulted in an increase of carrier mobility.

PANi/graphene composites fabricated by a combination of in situ polymerization and a solution process displayed a significant improvement in TE properties [41]. PANi chains were coated on the surface of graphene by the strong π-π conjugation interactions during in situ polymerization, and then the molecular structure of PANi switched coiled conformation to expanded chains by the chemical interactions between PANi and solution. Sequential chemical reactions of in situ polymerization and the solution process created the uniform dispersion of graphene in the PANi matrix and also generated numerous nano-interfaces that could enhance the carrier mobility of the composites. The poor dispersion of graphene due to strong van der Waals interactions between graphene sheets in the polymer matrix usually ends up with a decrease in the Seebeck coefficient, but this work showed that the in situ polymerization process enhanced the dispersion homogeneity of graphene. The enlarged contact surface area provided by uniformly dispersed graphene generated more graphene/PANi interfaces that boosted the energy filtering effect and strengthened the π electronic conjugation structure. The maximum electrical conductivity and Seebeck coefficient of the composite with 48 wt% graphene reached 814 S·cm^{-1} and 26 μV·K^{-1}, respectively, translating to a power factor of 55 μW·m^{-1}·K^{-2}, which is one of the highest values ever reported in the polymer/graphene composites.

PANi/graphite oxide (GO) prepared via in situ polymerization displayed that the GO played a role of a template where the aniline molecules were grown on the surface of GO, which induced an ordered structure of PANi with a high crystallinity during polymerization [42]. XRD, fourier-transform infrared spectroscopy (FTIR), and XPS confirmed that there existed a strong interaction between exfoliated GO and PANi, including hydrogen bonding, π-π stacking, and electrostatic interaction. The exfoliated GO possessed large surface area, which could render polymers to have strong interactions with GO and hence the PANi chains in the composite were further ordered due to the template effect of GO. PANi/GO exhibited the maximum electrical conductivity and Seebeck coefficient, each of which was up to 750 S·cm^{-1} and 28 μV·K^{-1}, respectively, and the ZT value reached 4.84×10^{-4}. Mitra et al. investigated the TE performances of the PANi/reduced graphene oxide (rGO) composites by varying its concentration of rGO and PANi [43]. It turned out that the ZT of PANi/rGO increased with an rGO of up to 50%. A uniform growth of PANi with an ordered structure with high crystallinity on the basal plane of the rGO sheets increased carrier mobility, as evidenced by a Hall Effect measurement. The PANi/rGO composites prepared by the in situ chemical oxidative polymerization of aniline exhibited the maximum ZT value of 0.0046, which was 5 times higher than that of pure PANi [105].

3.1.2. Poly(3,4-ethylenedioxythiophene)-poly(styrenesulfonate) (PEDOT:PSS)-Based Graphene Composites

PEDOT:PSS graphene composites with a different weight percentage of graphene, prepared by solution spin coating, exhibited an enhancement in TE properties [106]. The stacked graphene was dispersed by shear stresses imposed by the mixing and sonication process, which broke up the aggregated graphene. After dispersing the graphene, the PEDOT:PSS was intercalated into the graphene sheets and there existed strong π-π interactions between each other, as shown in Figure 4a. XPS clearly displayed that the peaks were shifted slightly towards higher binding energy levels, which indicated the electron donation from PEDOT:PSS to the graphene surface and subsequently increased the carrier concentration (Figure 4b). The strong π-π interactions between the PEDOT:PSS and the graphene surface helped to reduce conjugated defects in the polymer backbone because the PEDOT:PSS was tightly coated on the graphene surface. This, in turn, caused the conformational changes of PEDOT:PSS into linear or expanded-coiled molecular structures. The PEDOT:PSS thin films with 2 wt% graphene had the maximum power factor of 11 $\mu W \cdot m^{-1} \cdot K^{-2}$ at 300 K and the ZT value was up to 2.1×10^{-2}.

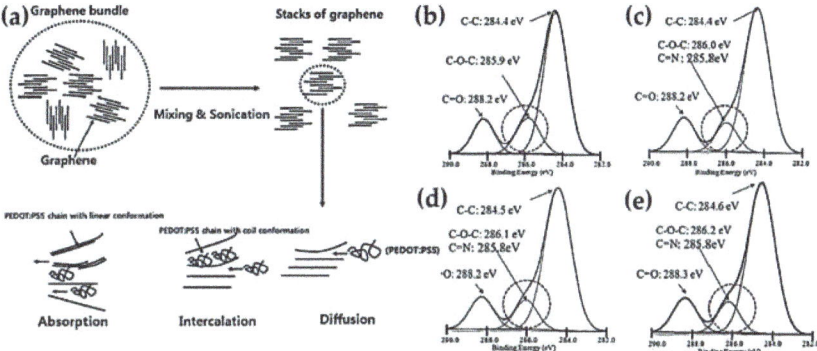

Figure 4. (a) The mechanism of dispersion of aggregated graphene in the polymer (PEDOT:PSS) during mixing and sonication processing. C1s x-ray photoelectron spectroscopy (XPS) peaks of (b) PEDOT:PSS, (c) PEDOT:PSS containing 1 wt% graphene, (d) PEDOT:PSS containing 2 wt% graphene, and (e) PEDOT:PSS containing 3 wt% graphene. Reprinted with permission from [106]. Copyright 2012 The Royal Society of Chemistry.

Doping with bromine can induce phonon scattering by introducing defects and decreasing the thermal conductivity. Ma et al. employed bromine doping into the PEDOT:PSS/graphene fiber to enhance TE performances [44]. Bromine doping made the Fermi level move down to the valence band due to the draining of electrons toward the highly electronegative Br sites. Lowering the Fermi level increased the density of the holes at the band edge, leading to an enhanced electrical conductivity and Seebeck coefficient. With an enhanced transport of charge carriers via bromine doping, the power factor was 624 $\mu W \cdot m^{-1} \cdot K^{-2}$ at room temperature.

PEDOT:PSS/rGO composites where reduced graphene oxide was dispersed in a PSS solution and followed by polymerization with the addition of EDOT monomer to the dispersion exhibited that the TE properties varied with the graphene content [45]. The electrical conductivity increased from 450 to 637 $S \cdot cm^{-1}$ at 3 wt% of graphene without the need for a reduction step such as the use of a toxic treatment process. Due to the enhanced electrical conductivity, the power factor of the PEDOT:PSS/rGO composites was up to 45.7 $\mu W \cdot m^{-1} \cdot K^{-2}$.

Noncovalently functionalized graphene with fullerene by π-π stacking in a liquid-liquid interface could help improve the TE properties of PEDOT:PSS. It was reported that fullerene could reduce the thermal conductivity via phonon scattering [107]. Zhang et al. introduced fullerene into the

PEDOT:PSS/graphene composites in which rGO was noncovalently functionalized with fullerene through π-π stacking in the liquid-liquid interface [108]. They found out that tailoring the fullerene and graphene ratio in the PEDOT:PSS matrix could tune the electronic and phonon transport and help increase the electrical conductivity due to the significant interfacial phonon scattering. The incorporation of fullerene/rGO nanohybrids enhanced the Seebeck coefficient as high as 4-fold relative to that of neat PEDOT:PSS film. Employing rGO into the conjugated polymer system could push the Fermi level away from the valence band, resulting in an increased Seebeck coefficient [109]. Zero-dimensional fullerene can allow quantum confinement where high energy carriers preferentially participate with the carrier transport while low energy carriers are impeded [56]. This further enhanced the Seebeck coefficient in the rGO-fullerene/PEDOT:PSS hybrid composites. The highest ZT reached 0.067 at 30 wt% nanohybrids-filled polymer composites, where the ratio of fullerene to graphene was 3:7.

PEDOT:PSS/rGO/fluorinated C_{60} (F-C_{60}) hybrid nanocomposites exhibited an enhanced power factor of 83.2 $\mu W \cdot m^{-1} \cdot K^{-2}$ [46]. The nanointerfaces created in between rGO, F-C_{60}, and PEDOT:PSS provided Schottky barriers, which hindered cold-energy carriers and preferentially allowed high energy carriers to pass through interfacial energy filtering. The synergistic combination of the rGO/F-C_{60} hybrid and PEDOT:PSS formed Schottky barrier due to work functions and resulted in a ZT of 0.1 at room temperature.

3.1.3. Other Conducting Polymers-Based Graphene Composites

Du et al. investigated the TE properties of graphene nanosheets (GN) in the poly(3-hexylthiophene) (P3HT) matrix, which was prepared by oxidative polymerization [47]. As the content of GN increased to 30 wt%, the electrical conductivity was optimized to 1.2 $S \cdot cm^{-1}$ and the Seebeck coefficient increased to 35.5 $\mu V \cdot K^{-1}$, resulting in the power factor of the P3HT/GN becoming 0.16 $\mu W \cdot m^{-1} \cdot K^{-2}$ at 30 wt% of graphene. A simultaneous increase in the electrical conductivity and Seebeck coefficient was ascribed to an increase in the carrier mobility of the composites with GN loading.

The incorporation of GN into the polypyrrole (PPy) matrix showed an enhanced TE property with a proper ratio of polymer to GN. PPy/GN composites, synthesized by a simple in situ chemical polymerization, exhibited that PPy grew along the surface of GN to form an ordered molecular structure with increased crystallinity [48]. The strong π-π interactions between PPy and GN created during polymerization induced the conformation of PPy molecular chains to change from a compacted coil to an expanded coil. Moreover, the GN was homogeneously dispersed in the PPy matrix and effectively bridged the carrier transport via the π-π interactions with PPy, which, in turn, improved the carrier mobility [110]. By increasing the GN content, the electrical conductivity and Seebeck coefficient enhanced simultaneously. The optimized power factor of the composites reached 10.2 $\mu W \cdot m^{-1} \cdot K^{-2}$, which is 250 times higher compared to that of pure PPy [16]. The thermal conductivity was measured to be 0.84 $W \cdot m^{-1} \cdot K^{-1}$ and therefore, the maximum ZT was up to 2.8 × 10^{-3}.

The three-dimensional (3D) interconnected nanocomposite where two-dimensional (2D) rGO was sandwiched by one-dimensional (1D) PPy nanowires has been synthesized via a convenient interfacial adsorption-soft template polymerization [49]. A 3D interconnected microstructure displayed that the PPy nanowires and the rGO nanolayers were connected and formed a 3D network architecture. During polymerization, the PPy nanowires were tightly attached on the surface of the rGO nanolayers, which provided conducting pathways for electron transport. At an rGO:PPy mass ratio of 50 wt%, the nanocomposite reached an electrical conductivity of 75 $S \cdot cm^{-1}$, which is about 60 times higher than that of the pure PPy nanowires, and the Seebeck coefficient was up to 34 $\mu V \cdot K^{-1}$, which translated to a power factor of 8.6 $\mu W \cdot m^{-1} \cdot K^{-2}$. This value is about 480 times higher relative to that of pure PPy nanowires.

Uniform PPy coating on rGO was fabricated via a template-directed in situ polymerization in which sodium dodecyl sulfate (SDS) was adsorbed on the rGO surface via van der Waals forces and helped disperse rGO [111]. SEM and TEM confirmed that exfoliated rGO flakes acted as the template in the core and PPy wrapped around the rGO as the shell layers. The exfoliated rGO nanosheets with the help of SDS was beneficial for the PPy molecules to have an ordered alignment on the rGO surface, which resulted in a great enhancement in both the electrical conductivity and Seebeck coefficient. The PPy/rGO composites with the rGO:PPy ratio of 2:1 exhibited a power factor of 3 µW·m^{-1}·K^{-2} at room temperature.

The ternary composites of PPy/GN/PANi synthesized by the combination of the in situ polymerization and solution process exhibited high TE performance, which stemmed from the strong π-π interactions between PPy, GN, and PANi through π-π stacking and hydrogen bonding, as shown in Figure 5a [50]. A uniform coating of PPy on the surface of graphene was achieved from in situ polymerizations, and mixing pure PANi with PPy/GN created the PPy/GN/PANi ternary composites. The highly ordered structure of polymers and the uniform dispersion of graphene in the polymer matrix during polymerization were very beneficial to create numerous interfaces between components, which may cause the increase in the energy filtering and the quantum-confinement effects and consequently enhance the carrier mobility [72]. A highly aligned polymer chain with an ordered molecular structure on the surface of the graphene nanosheets reduced the interchain and intrachain hopping barrier, which significantly lowered the carrier transport resistance. The TE properties were shown to be dramatically improved by combining three components compared to those of two component composites (PANi/GN and PPy/GN), as shown in Figure 5b. At 32 wt% of graphene, PPy/GN/PANi exhibited an electrical conductivity of 500 S·cm^{-1} and Seebeck coefficient of 32 µV·K^{-1}, leading to a maximum power factor of 52.5 µW·m^{-1}·K^{-2}.

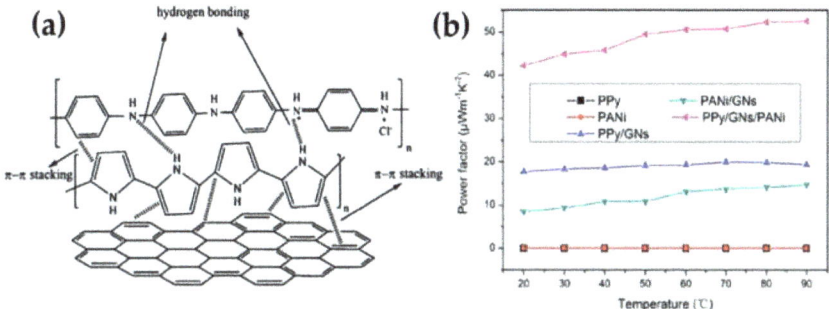

Figure 5. (a) The diagram of interactions of polypyrrole (PPy)/graphene nanosheets (GNs)/PANi composite and (b) Power factor of pure PPy, PANi, PPy/GNs composite, PANi/GNs composite, and PPy/GNs/PANi composite with 32 wt% graphene at different temperatures. Reprinted with permission from [50]. Copyright 2017 American Chemical Society.

3.2. Carbon Nanotube-Based Nanocomposites

3.2.1. Non-Conductive Polymers-Based Carbon Nanotubes (CNTs) Composites

Nafion, a water-soluble perfluorosulfonated polymer, has a hydrophobic nature in its backbone that could be used to solubilize CNTs in an aqueous solution. Choi et al. investigated the TE properties of Nafion/CNTs nanocomposites using the doctor blade method with different types of CNTs where single CNTs (SWNT), few CNTs (FWNT), and multi-walled CNTs (MWNT) were dispersed in an aqueous solution of Nafion [51]. For all CNTs studied, both the electrical conductivity and Seebeck coefficient increased with the concentration of CNTs. Although the Seebeck coefficient remained insensitive on the CNTs (ranged to 20–25 µV·K^{-1}), the electrical conductivity was dependent on the types of CNTs. Among the CNTs-based composites studied, the Nafion/MWNT films at the 30 wt% of MWNTs exhibited the highest electrical conductivity at 13 S·cm^{-1}. This work indicated that the electrical conductivity was dependent on the types of CNT and that the Seebeck coefficient was relatively insensitive of the CNT type, which indicated that high-energy-charges could participate in transport processes irrespective of the type of CNTs [62]. The maximum power factor was 1 µW·m^{-1}·K^{-2} in the Nafion/FWNT composites.

Segregated-network polymer/CNT composites investigated by Yu et al. exhibited that the electrical conductivity dramatically increased by creating a network of CNTs in the composite, while the Seebeck coefficient and thermal conductivity were independent upon the CNT concentration [52]. A poly(vinyl acetate) (PVAc) homopolymer emulsion, which is a stable suspension of solids, was used as a matrix for CNTs, and gum arabic (GA) was also used to stabilize the CNT in water. The 3D network of the CNTs was formed within the interstitial spaces between the emulsion particles where CNTs wrapped around the emulsion particles in a network fashion rather than being randomly distributed. The segregated network created thermally disconnected but electrically connected junctions in the nanotube network. With a CNT concentration of 20 wt%, the electrical conductivity and thermal conductivity of the composites were 48 $S \cdot cm^{-1}$ and 0.34 $W \cdot m^{-1} \cdot K^{-1}$, respectively, which translated to a ZT of 0.006 at room temperature. Since phonon and electron transport are affected by tube/tube junctions, controlling the carrier transport across the junctions by altering the stabilizer concentration influenced the TE behavior of latex-based polymer composites. Kim et al. investigated the TE behavior of PVAc/SWNT-GA composites by varying the SWNT:GA ratios [53]. The electrical conductivity was significantly increased from 5.27 to 90.4 $S \cdot cm^{-1}$ by changing the SWNT:GA ratio from 1:3 to 10:1 while the Seebeck coefficient and thermal conductivity showed a little change with varying SWNT:GA ratios, each of which remained 40 $\mu V \cdot K^{-1}$ and 0.25 $W \cdot m^{-1} \cdot K^{-1}$, respectively.

The same group further investigated the TE behaviors of the CNT-filled latex-based composites, where PVAc was used as a matrix for MWNT, stabilized with a semiconducting molecule, such as sodium deoxycholate (DOC) or meso-tetra(4-carboxyphenyl) (TCPP) [54]. Both surfactants induced the MWNT to be exfoliated in solution by adsorbing to their surfaces and changing them from hydrophobic to hydrophilic. Upon drying, the PAVc polymer chains excluded a volume that the MWNT could occupy, which forced the nanotubes into the interstitial spaces between them and resulted in a segregated MWNT network, as shown in Figure 6. The electrical conductivity of the PVAc/MWNT composites increased at relatively low MWNT concentrations for both stabilizers. The PVAc/MWNT-TCPP composites with 12 wt% MWNT exhibited an electrical conductivity of 1.28 $S \cdot cm^{-1}$ and a Seebeck coefficient of 28 $\mu V \cdot K^{-1}$. By replacing MWNT with double walled carbon nanotubes (DWNT), the PVAc/DWNT-TCPP composites revealed an electrical conductivity of 71.08 $S \cdot cm^{-1}$ and Seebeck coefficient of 78 $\mu V \cdot K^{-1}$, which translated to a power factor of 42.8 $\mu W \cdot m^{-1} \cdot K^{-2}$.

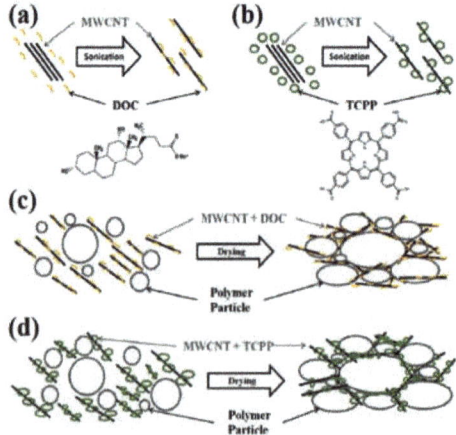

Figure 6. The dispersed multi-walled carbon nanotubes (MWNT), when mixed with polymer emulsion particles, forms a three-dimensional network in the interstitial positions between the polymer particles. (a,b) show representations of the dispersed MWNT in the two stabilizing agents, DOC and TCPP. (c,d) illustrate the formation of a segregated network upon the drying of the water-based polymer emulsions. Reprinted with permission from [54]. Copyright 2012 Elsevier.

3.2.2. PANi-Based CNTs Composites

Meng et al. demonstrated a novel approach to enhancing TE properties of the PANi/CNT nanocomposites prepared via a simple two-step method [56]. A freestanding CNT network consisted of individual CNTs and their bundles were fabricated by filtering a uniform CNT suspension. A PANi layer was coated on the CNT network via the in situ chemical polymerization method. By incorporating the CNT (1D materials) and PANi (quasi-1D materials) that exhibit larger TE properties than 2D or 3D ones due to a higher density of states at the Fermi level in low-dimensional structures, the resultant PANi/CNT composites enhanced the power factor to 5 $\mu W \cdot m^{-1} \cdot K^{-2}$ at 300 K [55,112]. PANi/SWNT nanocomposites prepared via a simple one-step in situ polymerization by using SWNTs as templates and aniline as reactant displayed the nanostructured PANi coating layer tightly wrapped around the SWNTs [60]. With increasing SWNT content, both the electrical conductivity and Seebeck coefficient of PANi/SWNT were enhanced. When the SWNT content was varied from 0 to 41.4 wt%, the electrical conductivity reached 125 $S \cdot cm^{-1}$ and the Seebeck coefficient increased from 11 to 40 $\mu V \cdot K^{-1}$. Such a dramatic improvement in the TE properties resulted from the increase of the carrier mobility in PANi that were aligned along the SWNTs with a good order of chain packing. The SWNTs further increased the carrier mobility by bridging the carrier transport through the strong π-π interactions with the rings of PANi. The maximum power factor at room temperature was up to 20 $\mu W \cdot m^{-1} \cdot K^{-2}$ for the 41.4 wt% SWNT and the thermal conductivity was 1.5 $W \cdot m^{-1} \cdot K^{-1}$, which translated to a maximum ZT of 0.004 at room temperature.

The PANi/CNT composites with an ordered molecular structure were also obtained by a combination of in situ polymerization and electron-spinning processes [58]. During in situ polymerization, aniline molecules were polymerized and aligned along the CNTs with a high degree of ordering due to the strong PANi and CNT interaction. Furthermore, the electrical field in the electro-spinning process caused PANi/CNT hybrids to align in a parallel manner into a long fiber. The highly ordered arrangement of polymer chains reduced the π-π conjugated defects and increased the effective degree of electron delocalization, which enhanced the carrier mobility in the composites [113]. The PANi/CNT nanofibers exhibited an electrical conductivity of 17 $S \cdot cm^{-1}$ and Seebeck coefficient of 10 $\mu V \cdot K^{-1}$ with CNT of 40 wt%. Zhang et al. prepared PANi/MWNT nanocomposites in a green method with no use of a dispersant [57]; that is, PANi and MWNTs were homogeneously mixed by cryogenic grinding (CG) and then the PANi/MWNT composites were consolidated by Spark Plasma Sintering (SPS). Upon SPS treatment, the electrical conductivity of the PANi/MWNT nanocomposites increased to 1.6 $S \cdot cm^{-1}$ by increasing the MWNT content from 10 to 30%, and the maximum power factor was 1.1 $\mu W \cdot m^{-1} \cdot K^{-2}$.

The in situ electrochemical polymerization process makes it possible for the polymer composites to adjust the morphology and microstructure. Liu et al. fabricated flexible PANi/SWNT composites and investigated the effect of the microstructure on their TE performance by modulating the electrolyte component and electric current [59]. During the electro-polymerization process, the conjugated structure of PANi chains was deposited onto the π-bonded surface of the SWNTs. PANi worked as the conductive glue to assemble the SWNTs into a homogeneous conductive network which facilitated the transfer of charge carriers. The SWNT network acted as a self-assembly template and accelerated the nucleation and the growth of PANi. SEM and TEM confirmed that more PANi was progressively electrodeposited onto SWNT with the increase in the diameter of PANi/SWNT composites. While the electro-polymerization proceeded, the composite films went through structural transitions to form 3-dimensional networks. The PANi/SWNT composites exhibited an electrical conductivity of 32 $S \cdot cm^{-1}$ and Seebeck coefficient of 45 $\mu V \cdot K^{-1}$ when the number of electro-polymerization reached 75. Further electrochemical polymerization made disordered PANi deposited, which resulted in the decrease in TE properties. The calculated power factor was up to 6.5 $\mu W \cdot m^{-1} \cdot K^{-2}$.

Regulating the degree of ordering of the molecular chain arrangements is an effective way to increase the electrical transport properties in organic TE materials. Wang et al. prepared PANi/SWNT nanocomposites by combining in situ polymerization and m-cresol solution processing [61]. During in situ polymerization SWNTs were well-dispersed in the nanocomposite and formed strong π-π interactions with PANi. The PANi/SWNT composite with SWNTs of 65 wt% exhibited an electrical conductivity of 1440 S·cm^{-1} and Seebeck coefficient of 39 μV·K^{-1}, which translated to a power factor of 217 μW·m^{-1}·K^{-2} at room temperature. This value was more than 20 times higher relative to that of the pure PANi film (Figure 7a,b). A large enhancement of TE properties in the nanocomposites was attributed to the highly ordered structure of PANi chains along the SWNTs via strong π-π interactions, which decreased the interchain and intrachain defects of the PANi molecules and consequently increased the carrier mobility. The Hall Effect measurement indicated that the carrier concentration slightly increased while the carrier mobility tripled with the SWNT contents. This result revealed that a simultaneous enhancement in both the electrical conductivity and Seebeck coefficient was mainly ascribed to the increase in carrier mobility. The thermal conductivity, measured by the laser flash technique at room temperature in the out-of-plane direction, was in the range of 0.2 to 0.5 W·m^{-1}·K^{-1} with different SWNT contents, as shown in Figure 7c. This low thermal conductivity could be explained by the fact that the numerous nano-interfaces formed in between PANi and SWNTs effectively scattered phonons and the differences in the vibrational spectra hindered phonon transport in the composites [114].

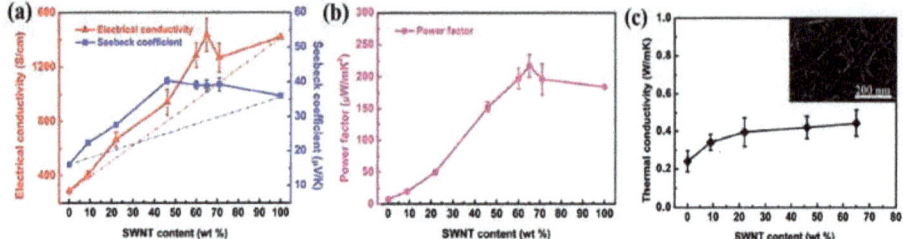

Figure 7. (a) The electrical conductivity and Seebeck coefficient, (b) power factor, and (c) thermal conductivity at room temperature of the PANi/single carbon nanotubes (SWNT) composites. The inset is the SEM image of PANi/SWNT composites with SWNT content of 65 wt%. Reprinted with permission from [61]. Copyright 2016 John Wiley and Sons.

3.2.3. PEDOT:PSS-Based CNTs Composites

The decoupled TE behaviors, which are ideal for realizing practical TE devices, between physical parameters such as electrical conductivity, Seebeck coefficient, and thermal conductivity can be achieved by modifying junctions between polymers and CNTs. A segregated network polymer-CNT composite created by drying water-based polymer emulsions that occurred after the addition of CNTs, stabilized in PEDOT:PSS, exhibited electrically connected, but thermally disconnected junctions [62]. With a high SWNT concentration of 35 wt%, the electrical conductivity was measured to be 400 S·cm^{-1}, but the Seebeck coefficient was relatively insensitive to the change, ranging from 15 to 30 μV·K^{-1}. The maximum power factor was up to 25 μW·m^{-1}·K^{-2}. The large improvement in electrical conductivity by raising SWNT loading did not affect the thermal conductivity of the DMSO doped-PEDOT:PSS/SWNT composites, which remained more or less the same in the range of 0.2–0.4 W·m^{-1}·K^{-1}. The TE properties of the segregated polymer-CNT network structure were enhanced by combining PEDOT:PSS, polyvinyl acetate, and SWNT [63]. Yu et al. investigated the TE behaviors of the polymer/SWNT composites consisting of SWNTs mixed with different grade PEDOT:PSS and/or PVAc polymers. Among samples investigated in their study, PEDOT:PSS (PH1000 type, 30 wt%)/PVAc (10 wt%)/SWNT (60 wt%) composites exhibited the highest TE performance with a power factor of 160 μW·m^{-1}·K^{-2}.

The same research group further improved the TE performance by using a dual-dispersants of TCC and PEDOT:PSS, which are intrinsically conductive and semiconducting stabilizers, respectively, for polymer/CNT composites [64]. In this work, two types of CNTs, including MWNT and DWNT, were exfoliated by using the combination of TCPP and PEDOT:PSS, both of which have been shown to stabilize CNTs through strong π-π interactions. Segregated-network polymer composites were created by embedding uniformly dispersed CNTs in TCPP and PEDOT:PSS into PVAc latex and followed by drying water-based polymer emulsions at an elevated temperature. As the water evaporated during drying, the polymer particles pushed the CNTs into interstitial spaces to form an electrically connected, but thermally disconnected 3D network. A composite made with PVAc latex and 40 wt% MWNTs 1:1:0.25 MWNT/PEDOT:PSS/TCPP exhibited 95 S·cm^{-1} (Figure 8a). The high electrical conductivity could be attributed to more favorable junctions between multiple stabilizing agents and CNTs in the segregated-network. An electrical conductivity of the sample system after replacing MWNTs with DWNTs was significantly improved by one order of magnitude (960 S·cm^{-1}) and the Seebeck coefficient was nearly doubled to 70 µV·K^{-1}, which translated to a power factor of 500 µW·m^{-1}·K^{-2} in the PVAc/PEDOT:PSS/TCPP/DWNT composites (Figure 8b–d).

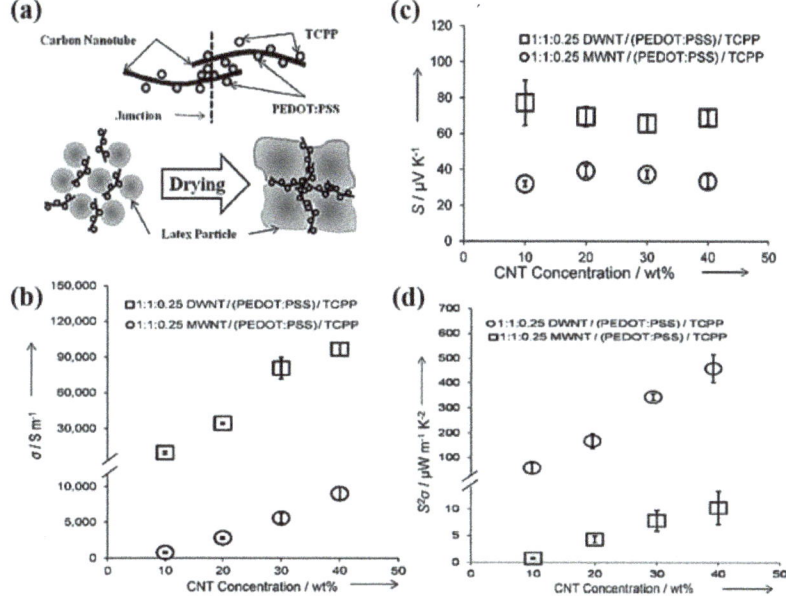

Figure 8. (a) top: the schematic of double walled carbon nanotubes (DWNT) coated by PEDOT:PSS and TCPP molecules and the junction formed between them. Bottom: schematic of the formation of a segregated network composite during polymer coalescence as it dries, (b) electrical conductivity, (c) seebeck coefficient, and (d) power factor as a function of CNT concentration and type. Reprinted with permission from [64]. Copyright 2013 John Wiley and Sons.

Zhang et al. reported that the TE properties of PEDOT:PSS/MWNT nanocomposites were greatly enhanced by a template-directed in situ polymerization [115]. Monomeric 3,4-ethylenedioxythiophene was adsorbed onto the MWNT that had been stabilized in PSS and polymerized after adding ammonium peroxydisulfate and iron chloride. Due to the strong π-π interaction and van der Waals interactions, PEDOT:PSS layers were effectively formed and coated on the surface of the MWNTs. At a ratio of PSS:EDOT = 0.3:1 and an MWNT:EDOT mass ratio of 70 wt%, the power factor of the PEDOT:PSS/MWNT nanocomposites reached a value of 0.23 µW·m^{-1}·K^{-2}.

The PEDOT:PSS/SWNT composites with a layered nanostructure were fabricated via two-step spin casting [65]. The double-layer nanostructure of the PEDOT:PSS/SWNT composite displayed a simultaneous improvement in electrical conductivity and Seebeck coefficient without the addition of dielectric solvents. The decoupled TE behavior was ascribed to the energy filtering effect. The layered nanostructure of PEDOT:PSS and SWNTs formed numerous nanometer-sized interfaces between each component that could create energy potential barriers. The maximum electrical conductivity and Seebeck coefficient of the PEDOT:PSS/SWNT composites reached 240 S·cm^{-1} and 39 µV·K^{-1}, respectively. The power factor was up to 21 µW·m^{-1}·K^{-2}, which is 4 orders of magnitude higher than the pure PEDOT:PSS.

Ethylene glycol (EG) treatment is an effective way to improve the electronic property of PEDOT:PSS-based composites by selectively removing non-conductive PSS chains out of the PEDOT:PSS films [31,116]. Lee et al. investigated the post-treatment of PEDOT:PSS/CNT nanocomposites by simply immersing the as-prepared films into the EG for 1 h, followed by annealing at 140 °C for 10 min to remove the residual EG [66]. Prior to post-treatment, the nanocomposite films with 25% DWNT exhibited a relatively low power factor of 38 µW·m^{-1}·K^{-2}. However, when subjected to EG treatment, the power factor of the nanocomposites dramatically increased up to 150 µW·m^{-1}·K^{-2}. The removal of the non-complexed PSS chains decreased the inter-CNT bundle distance and reduced the tunneling distance, consequently increasing the total carrier density of states and, hence, the electrical conductivity of the nanocomposite films.

3.2.4. Other Conducting Polymers-Based CNTs Composites

Other conducting polymers have been shown to be effective TE materials with high performances when compounded with carbon nanotubes. P3HT/SWNT nanocomposite films prepared by using simple-bar-coating exhibited a high TE performance without additional P3HT doping. Lee et al. investigated the TE properties of P3HT/SWNT films by optimizing the SWNT composition and solid content in the wire-bar-coating process [68]. Compared to drop-cast films, the wire-bar-coated P3HT/SWNT nanocomposite films, where SWNTs with diameters in the range of 6–23 nm formed well-dispersed and interconnected networks, exhibited a much higher TE performance with power factor up to 105 µW·m^{-1}·K^{-2} at room temperature for an ink with a total solid content of 4 mg mL^{-1}. Combining P3HT with MWNT via solution mixing showed a relatively high TE performance due to a uniformly dispersed MWNT network in the P3HT matrix [67]. The electrical conductivity of P3HT/MWNT increased remarkably from 0.02 to 30 S·cm^{-1} with an MWNT content from 10 to 90 wt%, and the Seebeck coefficient remained constant to the values between 23 and 28 µV·K^{-1}, while the thermal conductivity slightly increased to 0.59 W·m^{-1}·K^{-1}. More conductive pathways through the composites with increasing MWNT content improved the ZT value to 8.71×10^{-4} at 80 wt% of MWNT.

Bounioux et al. reported a high TE performance of the P3HT/SWNT composites that were compounded with SWNT followed by optimized p-doping [69]. Optimally p-doped P3HT/SWNT composites revealed an electrical conductivity of 1000 S·cm^{-1} and produced a power factor of 107 µW·m^{-1}·K^{-2} for 80 wt% SWNT. The TE properties can be dramatically enhanced by the sufficient doping of polymers in the hybrid films. Hong et al. doped P3HT chains with a 0.03 M FeCl$_3$/nitromethane solution using simple spin-coating onto the as-prepared P3HT/SWNT hybrid films [70]. During the doping process, dopants diffused into the polymer chains and captured electrons from them, which increased the electrical conductivity of the polymers while SWNT remained insensitive. The resultant P3HT/SWNT composites prepared by a simple bar-coating process exhibited an electrical conductivity of 2760 S·cm^{-1} and Seebeck coefficient of 31 µV·K^{-1}, producing a power factor of 308 µW·m^{-1}·K^{-2}, which tripled that of the hybrid films via conventional immersion method. Spray-printed P3HT/CNT nanocomposites exhibited an excellent TE performance [71]. The electrical conductivity and Seebeck coefficient were up to 348 S·cm^{-1} and 97 µV·K^{-1}, respectively, producing a power factor of 325 µW·m^{-1}·K^{-2} at room temperature.

Wang et al. fabricated PPy/MWNT composite powders by varying MWNT contents ranging from 0 to 20 wt% via an in-situ polymerization method using p-toluenesulfonic acid as a dopant and iron chloride as an oxidant [72]. The PPy/MWNT nanocomposites, composed of PPy nanoparticles and an MWNT core/PPY shell, exhibited that the electrical conductivity increased to 72 S·cm^{-1} with an MWNT of 15 wt%, which was more than 3 times as high as that of the pure PPy. An increase in TE properties was attributed to a conductive network where PPy chains were aligned onto MWNT with a more ordered crystalline structure via the π-π interaction between PPy and MWNT, which enhanced the carrier mobility. The PPy/MWNT nanocomposites at 20 wt% MWNT revealed a power factor of 2 $\mu W \cdot m^{-1} \cdot K^{-2}$ at room temperature, which was about 26 times higher relative than that of pure PPy.

The free-standing PPy/SWNT composite films, where nanosheets of SWNT, stabilized in SDBS, were physically mixed with PPy nanowires using the convenient procedure of solution mixing and subsequent common vacuum filtration, displayed a unique morphology with a high TE performance [73]. The pure PPy nanowires, synthesized by chemical oxidative polymerization using cetyltrimethyl bromide (CTAB) and ammonium peroxydisulfate (APS), exhibited an electrical conductivity of 2.2 $S \cdot cm^{-1}$ and a Seebeck coefficient of 10 $\mu V \cdot K^{-1}$, producing a power factor as low as 0.02 $\mu W \cdot m^{-1} \cdot K^{-2}$. PPy/SWNT composites with a layered morphology in which parallel SWNT nanosheets were sandwiched by PPy nanowires revealed an enhanced TE performance with the power factor as large as 21.7 $\mu W \cdot m^{-1} \cdot K^{-2}$, which was about 1000 times higher than that of the neat PPy nanowires.

3.3. Graphene/CNT-Based Nanocomposites

Hybridization by incorporating 1D and 2D carbonaceous nanofillers into a conducting polymer matrix has been shown to be an effective route for further improving TE properties via the synergistic effects from each component. Employing both low-dimensional carbon materials such as carbon nanotubes and graphene sheets into polymers has exhibited a dramatic enhancement in TE performance. For example, Yoo et al. fabricated a composite hybridized with graphene sheets and MWNT via the in situ polymerization of PEDOT:PSS in an aqueous solution [74]. The MWNT with a high aspect ratio (>1000) and graphene with a large surface area would provide long-range electrical bridges between conductive domains, which lowered the electrical transport resistance. The hybrid composite of PEDOT:PSS/graphene/MWNT with 5 wt% carbon materials exhibited an electrical conductivity and Seebeck coefficient of 690 $S \cdot cm^{-1}$ and 23 $\mu V \cdot K^{-1}$, respectively, which resulted in a power factor of 37 $\mu W \cdot m^{-1} \cdot K^{-2}$. Along with the thermal conductivity of 0.36 $W \cdot m^{-1} \cdot K^{-1}$, the ZT of the PEDOT:PSS/graphene/MWNT composite reached 0.31 at room temperature. The enhanced TE performance originated from the synergistic effects of multi-component systems with an excellent electrical bridging and electronic coupling between conductive domains.

Li et al. investigated the TE behaviors of polymer-based ternary composites which were synthesized by the in situ chemical oxidation of EDOT monomers on rGO and followed by physical mixing with SWNT [75]. Strong π-π interactions made the exfoliated SWNT adsorbed on the surfaces of the PEDOT/rGO composites. The resultant ternary composites of PEDOT/rGO/SWNT showed that the conducting networks consisting of 2D dispersed rGO nanosheets (surrounded with PEDOT polymers) and de-bundled SWNTs. Although the pure PEDOT had an electrical conductivity as low as 7.1 $S \cdot cm^{-1}$, the electrical conductivity of the PEDOT/rGO/SWNT composites was up to 38 $S \cdot cm^{-1}$. The post-treatment by H_2SO_4 further enhanced the electrical conductivity up to as high as 208 $S \cdot cm^{-1}$. With a slight decrease in Seebeck coefficient after the post-treatment, the ternary composites displayed increases in the power factors with increasing SWNT contents. The H_2SO_4-treated PEDOT/rGO/SWNT composites at 10% SWNT revealed a power factor of 9 $\mu W \cdot m^{-1} \cdot K^{-2}$.

The hybridization of carbonaceous nanofillers and conducting polymers by using layer-by-layer (LbL) was reported as a unique way to fabricate nanostructured composites with great potential to improve the TE performances [81,117]. A key advantage of the LbL technique is that of the sequential assembly of intrinsically conductive polymer and carbonaceous nanoparticle networks that leads to a novel three-dimensional film whose TE properties exceed those of each individual component, as well as those of a bulk film made with the same components. For instance, Cho et al. fabricated PANi-based carbon nanocomposites via LbL assembly where DWNT and graphene were stabilized with sodium dodecylbenzenesulfonate (SDBS) and poly(4-stryrenesulfonic acid) (PSS), respectively [76]. AFM on the surface of the PANi/graphene/PANi/DWNT thin films displayed that DWNTs were interwoven with each other, creating nanotube bundles due to their high concentration (Figure 9a). The addition of surfactants allowed DWNT and graphene to be exfoliated and uniformly dispersed in the LbL thin films, forming a 3D network with polymer-like entanglements of nanotubes and graphene platelets. SEM images in Figure 9b showed that individual DWNT and their bundles randomly intertwined together to form a well-dispersed network. TEM in Figure 9c exhibited that two overlapped graphene sheets (average diameter 1–1.5 µm) are surrounded by a PANi–DWNT matrix, where the PANi and DWNT were an interconnected network. Ordered PANi/graphene/PANi/DWNT nanocomposites with a uniform alignment of the 3D network structure yielded a power factor of 1825 $\mu W \cdot m^{-1} \cdot K^{-2}$. By fully taking advantage of the nanoscale engineering using the LbL approach, the PANi, graphene, and DWNT in the QL films exhibited a strong synergistic effect that surpassed the bilayer (BL)-LbL nanocomposites (PANi/graphene and PANi/DWNT), without sacrificing the intrinsic electrical properties of the individual carbon components (Figure 9d–g).

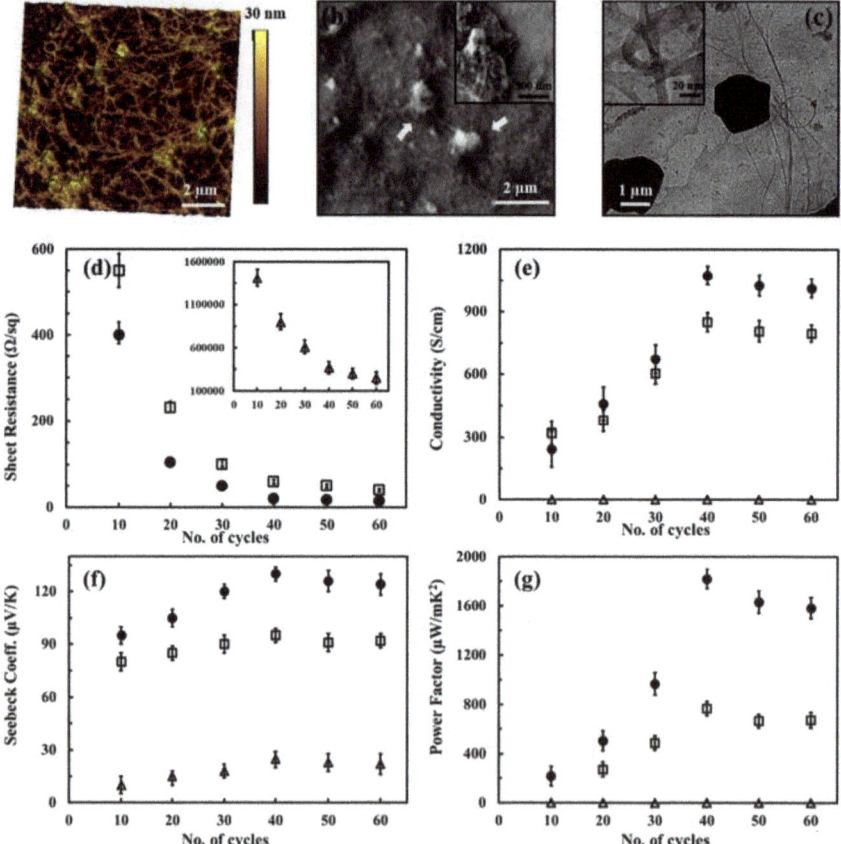

Figure 9. (a) The AFM 3D height image, (b) SEM, and (c) TEM images of the thin films. The insets in the SEM and TEM images show higher resolution graphene platelets and DWNT, respectively. The arrows indicate graphene in the film. (d) Sheet resistance, (e) electrical conductivity, (f) seebeck coefficient, and (g) power factor of PANi/graphene (open triangles), PANi/DWNT (open squares), and PANi/graphene/PANi/DWNT (filled circles) as a function of bilayers or quadlayers (i.e., cycles) deposited on a poly(ethylene terephthalate) (PET) substrate. Reprinted with permission from [76]. Copyright 2015 John Wiley and Sons.

By replacing all insulating stabilizers (PSS and SDBS) with water-soluble (or dispersible) intrinsically conductive polymers, PEDOT:PSS, an 80 quadlayer thin film ($\approx$1 µm thick), comprised of a PANi/graphene-PEDOT:PSS/PANi/DWNT-PEDOT:PSS repeating sequence, exhibited an unprecedented electrical conductivity of 1.9×10^3 S·cm^{-1} and a Seebeck coefficient of 120 µV·K^{-1} [77] These two values yielded a power factor of 2710 µW·m^{-1}·K^{-2}.

3.4. N-Type Thermoelectric Nanocomposites

In order to achieve practical applications, it is simultaneously required to fabricate both efficient p-type and n-type materials with a high TE performance. Recently significant progress in p-type materials has resulted in a high TE performance, however, most n-type counterparts have lower TE properties owing, in a large part, to difficulties in stable doping of organic materials and the lack of efficient n-type doped materials [81,118]. The large differences in the power factor between p- and n-type materials limit the widespread use of organic TE devices because of the unbalanced transport coefficients, which leads to power losses in π-leg module structures [119]. Semiconducting nanotubes are intrinsically of the n-type, but converted into p-type in the air due to their susceptibility on oxygen doping [120]. That is, n-type semiconducting organic materials such as carbon nanotubes (CNTs) are susceptible to oxygen doping in the air and the Seebeck coefficient becomes positive over time due to absorbed O_2 molecules that withdraw ~1/10 of an electron from CNTs [120–122]. To utilize the full potential of TE devices, it is critical to developing n-type polymers and composites to pair with their p-type counterparts. Towards addressing a lack of high-performance n-type characteristics of organic materials, great efforts have been paid to develop new strategies to fabricate n-type TE materials having excellent properties along with a high stability in the air [123–125].

All-polymer films have been attempted to produce air-stable n-type TE materials. For example, Sun et al. reported powder-pressed n-type TE films, assembled with conducting polymers poly(K_x(Ni-1,1,2,2-ethenetetrathiolate)s) (poly(Ni-ett)s), achieving a stable high power factor of 66 µW·m^{-1}·K^{-2} and a ZT value of 0.2 [78]. However, these polymers are neither soluble nor fusible, which significantly limits their processability. CNTs have been shown to be effectively converted into n-type TE performances through various methods. Solution-processed air-stable n-type TE materials were developed by synthesizing cobaltocene-encapsulated SWNTs [79]. The SWNT film doped with cobaltocene as an n-type showed a large power factor (75.4 µW·m^{-1}·K^{-2}) and low thermal conductivity of 0.15 W·m^{-1}·K^{-1}, reaching a ZT value of 0.157 at 320 K. Nonoguchi et al. demonstrated stable and efficient n-type SWNT doping with a series of salts such as sodium chloride, sodium hydroxide, and potassium hydroxide with crown ethers [126]. Their new n-type TE materials exhibited a ZT of 0.1 with an unprecedented air stability even at 100 °C for more than 30 days.

A high TE property of n-type CNTs-based composites has been obtained by decorating them with organic such as polyethylenimine (PEI) [127]. The donation of electrons from PEI to DWNTs effectively converted p-type nanotubes into n-type PEI/DWNT composites. After decorating DWNT with PEI through a stirring process, the Seebeck coefficient became −58 µV·K^{-1}. Freeman et al. improved an n-type behavior of CNTs-filled PEI composites [128]. They dispersed CNTs in SDBS under a tip sonication process and added 5 wt% of PEI solution into CNTs solutions in which PEI was expected to be attached on the surface of CNTs by physisorption. The PEI-coated CNTs were then made into composites with polyvinyl acetate (PVAc). The resultant composites exhibited an electrical conductivity of 15 S·cm^{-1} and Seebeck coefficient as large as −100 µV·K^{-1}. A high n-type behavior was believed to be due to the increase in the number of tubes that were evenly coated with PEI in well-dispersed CNTs in SDBS. The physical adsorption of PEI on the nanotubes made CNTs electron-rich due to the electron transfer from the amine groups in PEI. The lone pairs of the amine groups in PEI effectively donated electrons into the CNTs, resulting in an upward shift of the Fermi energy compared to an initial energy state and thus converting p-type CNTs into n-type. The n-type TE behaviors based on CNTs and PEI were further enhanced by the same group [80]. After functionalizing the CNTs with PEI, sodium borohydride ($NaBH_4$) was used as an n-type doping agent for CNTs. $NaBH_4$ could dope the CNTs where there were left undoped by PEI, which further improved the n-type characteristics. Therefore, the combination of PEI and NaBH$_4$ doping resulted in more effective n-type doping relative to the PEI-doped counterparts. The electrical conductivity of the composites was as high as 240 S·cm^{-1} and the Seebeck coefficient was −80 µV·K^{-1}.

A high power factor of graphene and carbon nanotubes polymer composites was achieved by using an LbL method in which DWNT-PEI/graphene-PVP were fabricated by alternately depositing DWNT, stabilized by PEI, and graphene stabilized by polyvinylpyrrolidone (PVP), from water, as shown in Figure 10a [81]. The electrical conductivity increased in an absolute form with the number of layers deposited, which suggests an increase in the density of the intersecting pathways for electron transport as the conductive network formed by the nanotubes; the graphene transitioned from 2D to 3D with the increasing layers (Figure 10b). The nanocomposites exhibited a modest increase in the Seebeck coefficient with thickness, attaining 80 µV· K^{-1} (Figure 10c). Both PEI and PVP contain nitrogen atoms either in the backbone or in the side group of the polymer chains, which allows them to act as electron donors. An 80-bilayer DWNT-PEI/graphene-PVP thin film (~320 nm in thickness) revealed a power factor of 190 µW·m^{-1}·K^{-2} at room temperature. Furthermore, unlike most organic n-type materials, this unique nanocomposite was relatively air-stable, which was demonstrated by testing after 60 days with no protection against moisture and oxygen (Figure 10d). An air-stable TE thin film assembly was believed to be due to the fact that LbL deposition produced nanocomposites of highly aligned and exfoliated graphene layers that created an extreme tortuosity for gas diffusion.

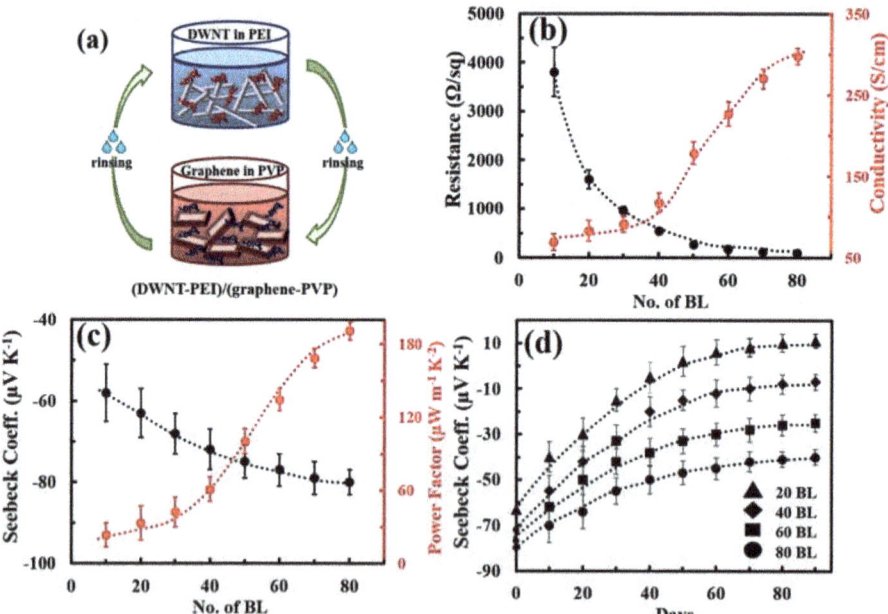

Figure 10. (a) The schematic of the layer-by-layer deposition process and molecular structures of the materials used. (b) Sheet resistance and electrical conductivity and (c) Seebeck coefficient and power factor, and (d) air-stability of Seebeck coefficient under ambient conditions. Reprinted with permission from [81]. Copyright 2016 Elsevier.

The addition of polyethylene glycol (PEG) into SWNT-based polymer composites is an effective way to convert p-type to n-type TE materials. Luo et al. prepared polypropylene (PP)/SWNT composites through a melting process [82]. By increasing the SWNT content from 0.8 to 2.0 wt%, the electrical conductivity of the PP/SWNT composites increased. The addition of a p-type nanoparticle, CuO, improved the electrical conductivity due to more charge carriers injected into composites. Interestingly, upon the addition of 10 wt% PEG into 2 wt% SWNTs and 5 wt% CuO composites, the Seebeck coefficient was changed from 37 to −57 µV· K^{-1}, and the power factor of the PP/SWNT/CuO/PEG composites increased to 7.8 × 10^{-2} µW·m^{-1}·K^{-2}.

A dramatic improvement in the power factor of the n-type organic materials has been recently made by judiciously controlling the electronic structure of spun carbon nanotube webs using various molecular dopants [83]. An et al. investigated the TE behaviors of a CNT web through the doping characteristics of n-type dopants and annealing treatment to remove oxygen from the surfaces of CNTs. To further improve the n-type TE performance, a benzyl viologen (BV) molecular dopant with the lowest reduction potentials among n-type organic molecules was used instead of an amine-rich polymer, PEI. The use of BV made the Fermi energy level of the CNT web further shift upward to the conduction band. Annealing the p-type CNT web at 300 °C for 10 h, followed by immersion into 2 mg mL^{-1} of BV solution for 8 h, increased the electrical conductivity to as high as 2220 S·cm^{-1} and the Seebeck coefficient up to −116 µV· K^{-1}. The annealed CNT web with BV treatment exhibited a power factor of 3100 µW·m^{-1}·K^{-2}, which is one of the highest values ever reported in organic composites.

4. Flexible Organic Modules

Thermoelectric generators (TEGs) have been recognized as a useful waste heat recovery system due to their simple structure, high power density, and lack of noise pollution. Typically, TEGs consist of legs of alternating p-type and n-type materials, each of which is connected electrically in series and thermally in parallel. The maximum heat to electrical energy conversion efficiency (η_{max}) of a TE material depends on its Carnot efficiency and TE properties. In TEGs, the efficiency of η_{max} is expressed as [129]

$$\eta_{max} = \frac{T_h - T_c}{T_h} \frac{\sqrt{1+ZT_m} - 1}{\sqrt{1+ZT_m} + \frac{T_c}{T_h}}$$

where $T_m = (T_h + T_c)/2$, and T_c and T_h are the cold-side and hot-side temperatures of the device, respectively. Although traditional inorganic-based bulk TE materials have achieved ZT values over 2 along with high power output and power density, this high value is measured at temperatures above 600 K. The ZT of these materials drops below 0.5 with a power factor <400 µW·m^{-1}·K^{-2} at room temperature [130]. In other words, most inorganic TE materials generate a greater amount of electricity at elevated temperatures (>500 K) [129]. This has restricted their power generation applications for energy harvesting from a variety of heat sources [15]. Considering that more than 50% of waste heat is stored at temperatures <500 K, the use of organic TE materials for the efficient harvesting of waste heat in the 300–500 K range could offer versatile power generation by fully capturing low-graded heat [131]. It is apparent that there will be a growing demand to develop flexible TEGs that could be attached to the device or on curved surfaces like human bodies with recent advances in flexible and stretchable electronic devices. For example, harvesting the heat dissipated from the human body could charge the batteries of small electronics or run low-power devices including medical sensors and wristwatches simply by embedding TEGs into an article of clothing. In this regard, the organic TE nanocomposites characterized by mechanical flexibility and the ready availability of such heat sources in daily life are suitable for developing wearable power generators. The proper utilization of the abundant low-grade heat waste through organic TEGs is a promising technology for various niche applications.

All polymer-based p- and n-type TE devices showed high power output performance. Sun et al. fabricated an inkjet-printed flexible device (composed of p- and n-type composites) that was obtained by ball-milling the poly[Ax(M-ett)] (A = Na, K; M = Ni, Cu) with a poly(vinylidene fluoride) solution. An all-polymer TE module consisting of six thermocouples that were printed onto the PET substrate produced a maximum output voltage of 15 mV and a maximum output power of 45 nW with a load resistance of 5 KΩ upon the application of a 25 K temperature gradient. The same research group developed high-performance organic TEGs by combining n-type poly[Nax(Ni-ett)] and p-type poly[Cux(Cu-ett)] [78]. After both polymers were compressed into cuboids, the module composed of 35 p-n couples on an AlN substrate was fabricated with silver and Al to interconnect the bottom and top substrates (Figure 11a,b). The output voltage and short-circuit current increased steadily with temperature, producing an open voltage of 0.26 V and a current of 10.1 mA at $\Delta T = 80$ K (Figure 11c). A maximum output power of 750 µW was generated with a load resistance of 33 Ω under a temperature difference of $\Delta T = 82$ K, which is the highest power ever reported in organic TEGs (Figure 11d). This high TE performance was attributed to the excellent TE properties of both p-type (ZT~0.01) and n-type (ZT~0.1–0.2 around 400 K) conducting polymers in the module.

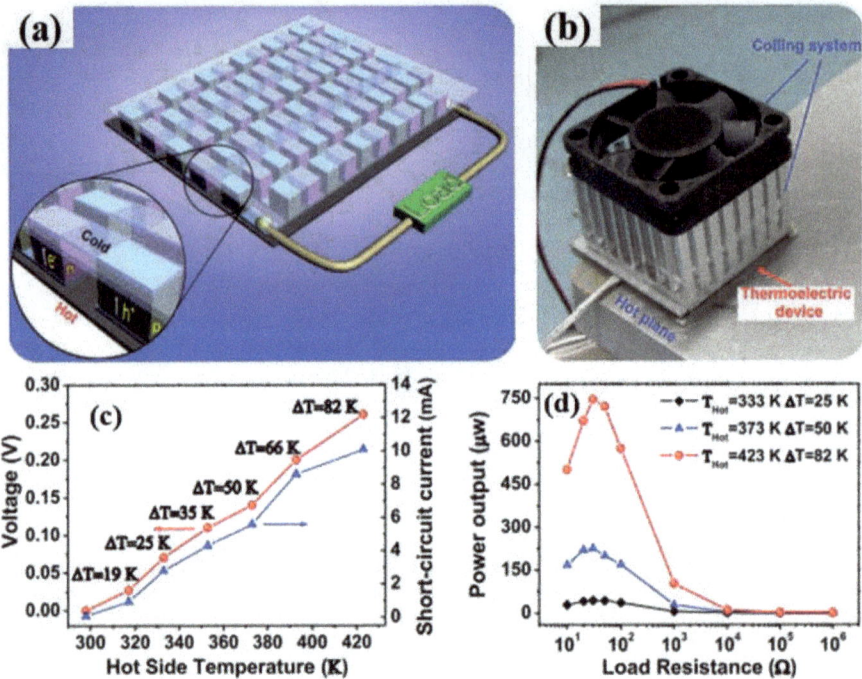

Figure 11. A thermoelectric module consisting of 35 thermocouples. (**a**) Module structure. (**b**) Photograph of the module and the measurement system with a hot plane and cooling fan. (**c**) The output voltage and short-circuit current at various Thot and ΔT. (**d**) The measured power output of the module with different loads. Reprinted with permission from [78]. Copyright 2012 John Wiley and Sons.

Crispin et al. fabricated thermoelectric generators (TEGs) using PEDOT:Tos as a thermoelectric p-type material and TTF-TCNQ as an n-type semiconductor. The maximum power output per area of the PEDOT-Tos/TTF-TCNQ TEGs, consisting of 54 thermocouples with the leg dimensions (25 mm × 25 mm × 30 mm), produced 45 µW·cm^{-2}. Wei et al. reported on organic thermoelectric modules screen-printed on paper by using PEDOT:PSS and silver paste, as shown in Figure 12 [132]. They used a highly conductive PEDOT:PSS solutions as an ink to cover a sheet of paper. Silver paste, having the good wettability on PEDOT:PSS, was screen-printed on the PEDOT:PSS legs to create either series (Figure 12a) or parallel (Figure 12b) connections. The large-area devices in which the PEDOT/silver paste arrays were sandwiched between Cu plates (Figure 12c,d) and connected with metal wires either in series or in parallel generated a power output of 50 µW with a temperature gradient of 100 K.

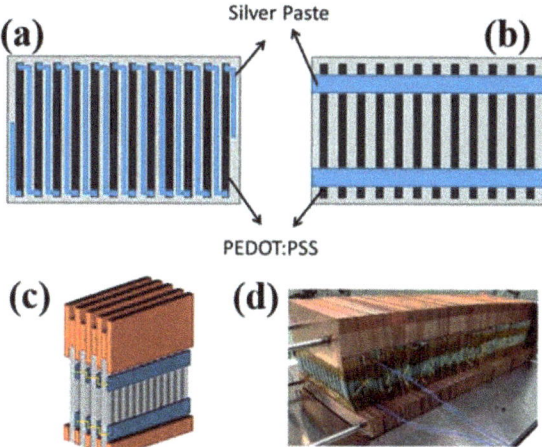

Figure 12. The schematic representation of the (**a**) series and (**b**) parallel PEDOT:PSS arrays; (**c**) the schematic and (**d**) photograph of the PEDOT:PSS modules sandwiched between copper plates. Reprinted with permission from [132]. Copyright 2014 The Royal Society of Chemistry.

Hybrid nanomaterials in which polymers are combined with carbon nanofillers such as graphene and carbon nanotubes have proven to be promising TE materials in fully utilizing low-graded heat for niche applications. For instance, Dörling et al. demonstrated the TE behavior of polymer composites where an initially p-type polymer/CNT composite was switched to n-type via UV photoinduction [133]. They fabricated flexible TEGs from solution using a single processing step; nitrogen-doped CNTs were dispersed in o-dichlorobenzene and dissolved in P3HT, followed by their mixing, sonicating, and drop-casting on PET substrates. The UV irradiation on the p-type P3HT/MWNT converted it to an n-type composite. A total of 15 double legs of p- and n-type P3HT/MWNT materials, which were attached to a glass filled with ice water, generated a voltage of 5 mV, showing the possibility for application as a wristband.

Wang el al. reported a generator using diethylenetriamine (DETA) as an n-type dopant for SWNTs [134]. They reported TEGs, composed of 14 thermocouples, with a high power output of 649 nW at $\Delta T = 55$ K. Other significant work was published in 2016 by Cho et al. in which the TEGs were fabricated with PANi/Au-doped-CNT (p-type TE materials) and PEI/CNT (n-type TE materials) as TE elements [135]. They achieved 376 nW using 14 thermocouple modules by applying a temperature gradient of 10 K. A unique combination of the layer-by-layer assembly of multiwalled carbon nanotubes and in-situ electrochemical polymerization of PEDOT exhibited high TE properties [136]. The MWNT films (<1 µm thick) infused with PEDOT displayed a power factor of 155 µW/m K^2. A cylindrical TE module using p-type TE materials of poly(diallyldimethylammonium chloride) (PDDA)/MWNT/PEDOT produced a maximum power of 5.5 µW with a temperature gradient of 30 K.

Recently, textile thermoelectric generators have become very popular in the field of organic TEG due to their easy integration in fabrics for body heat energy recovery. In terms of harvesting low-grade heat generated from the human body, a fabric-based TE material is suitable for wearable self-powered electronic devices with the merit of its flexibility and wearability, which is very difficult to achieve with conventional inorganic-based counterparts. Wu et al. fabricated yarns with a TE functionality in which the yarns were processed with waterborne polyurethane (WPU), MWNT, and PEDOT:PSS composites [137]. Nonionic WPU solution was used as a viscous polymer matrix, and MWNT as a conductive filler was stabilized in PEDOT:PSS in water. The synthesized nonionic WPU was completely dissolved in water because of the hydrophilicity of the PEG in WPU. By increasing the ratio of PEDOT:PSS in MWNT solutions, the electrical conductivity and Seebeck coefficient of the WPU/MWNT-PEDOT:PSS composites increased simultaneously, leading to the enhancement of power factors. At the ratio of 1:4 of MWNT to PEDOT:PSS and doping with 5 wt% DMSO, the composites exhibited a power factor of 1.4 $\mu W \cdot m^{-1} \cdot K^{-2}$. Based on the optimized conditions (20 wt% MWNT 1:4 ratios of MWNT to PEDOT:PSS, and doped with 5 wt% DMSO), they applied to typical yarns such as cotton and polyester as the substrate to evaluate the coating feasibility. The resistance of both yarns decreased with the number of dipping cycles, and polyester yarn displayed a lower resistance than that of the cotton, which may stem from a uniform and smooth surface in long continuous filament fibers of the polyester yarn, while the cotton yarn had short fibers with uneven and rough surfaces.

Du et al. made flexible fabric-based TE devices by incorporating PEDOT:PSS onto commercial fabric [138]. The PEDOT:PSS coated fabric maintained the flexibility and softness of the polyester fabric, which displayed a high mechanical compliance upon deformation such as rolling, bending, and twisting. The PEDOT:PSS coated fiber exhibited a power factor of 0.045 $\mu W \cdot m^{-1} \cdot K^{-2}$. The electric voltage generated by fabric-based PEDOT:PSS devices increased in a linear relationship with both temperature differences and the number of strips placed in series. The modules of the fabric-based TE generator connected with 5 strips produced 4.3 mV with a ΔT of 75.2 K.

A novel design of flexible TE generators, composed of n-type yarns, were fabricated in a way that commercial poly(ethylene terephthalate) (PET) sewing threads were coated with poly(N-vinylpyrrolidine) (PVP)/MWNT nanocomposites [139]. The n-type yarns exhibited an electrical conductivity of 1 S·cm^{-1} and Seebeck coefficient of −14 $\mu V \cdot K^{-1}$, which was stable for several months at ambient conditions. A textile TE module of 38 embroidered n/p strips, consisting of the combination of n-type yarns with p-type PEDOT:PSS coated silk yarns, produced an open-circuit voltage of 143 mV with a temperature difference of 116 K and a maximum power output was as high as 7.1 nW at a temperature gradient of 80 K, as shown in Figure 13.

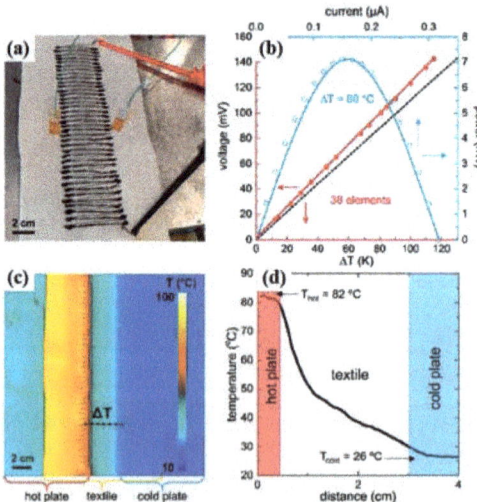

Figure 13. (a) The all-organic in-plane embroidered textile thermoelectric (TE) device with 38 n/p elements—constructed with n-type coated PET yarns, p-type dyed silk yarns and a conducting carbon-based paste for electrical connections. (b) Electrical measurements of the module with a measured output voltage V_{out} as a function of ΔT (red line), and calculated (dotted line), as well as power output P = V_{out} I as a function of measured current I for ΔT ≈ 80 K. (c) Thermal image of the module, placed as in (a), with T_{hot} ≈ 82 K and T_{cold} ≈ 26 K. (d) Temperature gradient across the textile device. Reprinted with permission from [139]. Copyright 2018 American Chemical Society. Further permissions related to this article should be directed to the American Chemical Society.

5. Summary and Outlook

There is no doubt that in the last few years organic thermoelectric materials have become in a real alternative (in terms of power factor) to the traditional inorganic semiconductors used in the commercial manufacture of thermoelectric generators. Since 2008, the values of the power factors of organic thermoelectric materials have been increased by 4 orders of magnitude while the improvement of inorganic semiconductors has remained in the same order. The possibility of tuning the doping level of organic semiconductors and the control of the polymer chain conformation have been a key to achieving the enormous development of the thermoelectric efficiency of conducting polymers and their corresponding organic composites. Polymer nanocomposites compounded with carbon nanofillers have been shown to rival thermoelectric efficiency in terms of power factor with inorganic-based TE materials; for example, multilayered systems such as PANi/graphene-PEDOT:PSS/PANi/DWNT-PEDOT:PSS (PF ~ 2710 $\mu W \cdot m^{-1} \cdot K^{-2}$) or the annealed CNT web with BV treatment (PF ~ 3100 $\mu W \cdot m^{-1} \cdot K^{-2}$). In addition, carbon materials can behave as n-type semiconductors through doping with molecules or polymers having electron donor groups (such as amine groups). However, there are several challenges to address in the field of organic semiconductors for practical thermoelectric applications. One of the main limitations of organic TE materials is that they cannot be used for high-temperature applications due to their degradation problems. Thus, the temperature gradient across the thermoelectric generator is very limited and the possible way of producing a considerable amount of energy is their use in larger scale areas. For this reason, a future research effort will be needed to find a cost-effective way for producing organic thermoelectric nanocomposites for large-scale areas and to develop a promising technology for their commercial production beyond their simple niche applications. In addition, future investigations should be focused on the integration of organic thermoelectric generators into fabrics in order to harness low-graded energy sources such as body heat, which could be utilized to power small sensors and wireless devices in an environmentally friendly way.

Funding: This research received no external funding.

Acknowledgments: This research was supported by Wonkwang University in 2018.

Conflicts of Interest: The authors declare no conflict of interest.

References

1. Edenhofer, O.; Pichs-Madruga, R.; Sokona, Y.; Seyboth, K.; Matschoss, P.; Kadner, S.; Zwickel, T.; Eickemeier, P.; Hansen, G.; Schlömer, S. IPCC special report on renewable energy sources and climate change mitigation. In *Prepared By Working Group III of the Intergovernmental Panel on Climate Change*; Cambridge University Press: Cambridge, UK, 2011.
2. Panwar, N.; Kaushik, S.; Kothari, S. Role of renewable energy sources in environmental protection: A review. *Renew. Sustain. Energy Rev.* **2011**, *15*, 1513–1524. [CrossRef]
3. Abolhosseini, S.; Heshmati, A.; Altmann, J. *A Review of Renewable Energy Supply and Energy Efficiency Technologies*; SSRN: Rochester, NY, USA, 2013.
4. Forman, C.; Muritala, I.K.; Pardemann, R.; Meyer, B. Estimating the global waste heat potential. *Renew. Sustain. Energy Rev.* **2016**, *57*, 1568–1579. [CrossRef]
5. Cullen, J.M.; Allwood, J.M. Theoretical efficiency limits for energy conversion devices. *Energy* **2010**, *35*, 2059–2069. [CrossRef]
6. DoE, U. *Waste Heat Recovery: Technology and Opportunities in US Industry*; US Department of Energy Industrial Technologies Program: Washington, DC, USA, 2008.
7. Zhang, Q.; Sun, Y.; Xu, W.; Zhu, D. Organic thermoelectric materials: Emerging green energy materials converting heat to electricity directly and efficiently. *Adv. Mater.* **2014**, *26*, 6829–6851. [CrossRef] [PubMed]
8. Weathers, A.; Khan, Z.U.; Brooke, R.; Evans, D.; Pettes, M.T.; Andreasen, J.W.; Crispin, X.; Shi, L. Significant Electronic Thermal Transport in the Conducting Polymer Poly(3,4-ethylenedioxythiophene). *Adv. Mater.* **2015**, *27*, 2101–2106. [CrossRef] [PubMed]

9. Vashaee, D.; Shakouri, A. Improved thermoelectric power factor in metal-based superlattices. *Phys. Rev. Lett.* **2004**, *92*, 106103. [CrossRef] [PubMed]
10. Biswas, K.; He, J.; Blum, I.D.; Wu, C.-I.; Hogan, T.P.; Seidman, D.N.; Dravid, V.P.; Kanatzidis, M.G. High-performance bulk thermoelectrics with all-scale hierarchical architectures. *Nature* **2012**, *489*, 414. [CrossRef] [PubMed]
11. Bahk, J.-H.; Bian, Z.; Shakouri, A. Electron energy filtering by a nonplanar potential to enhance the thermoelectric power factor in bulk materials. *Phys. Rev. B* **2013**, *87*, 075204. [CrossRef]
12. Zhang, X.; Zhao, L.-D. Thermoelectric materials: Energy conversion between heat and electricity. *J. Mater.* **2015**, *1*, 92–105. [CrossRef]
13. Yazdani, S.; Pettes, M.T. Nanoscale self-assembly of thermoelectric materials: A review of chemistry-based approaches. *Nanotechnology* **2018**, *29*, 432001. [CrossRef] [PubMed]
14. Russ, B.; Glaudell, A.; Urban, J.J.; Chabinyc, M.L.; Segalman, R.A. Organic thermoelectric materials for energy harvesting and temperature control. *Nat. Rev. Mater.* **2016**, *1*, 16050. [CrossRef]
15. LeBlanc, S. Thermoelectric generators: Linking material properties and systems engineering for waste heat recovery applications. *Sustain. Mater. Technol.* **2014**, *1*, 26–35. [CrossRef]
16. Yue, R.; Xu, J. Poly(3,4-ethylenedioxythiophene) as promising organic thermoelectric materials: A mini-review. *Synth. Metals* **2012**, *162*, 912–917. [CrossRef]
17. Chen, G.; Xu, W.; Zhu, D. Recent advances in organic polymer thermoelectric composites. *J. Mater. Chem. C* **2017**, *5*, 4350–4360. [CrossRef]
18. Liu, J.; Wang, X.; Li, D.; Coates, N.E.; Segalman, R.A.; Cahill, D.G. Thermal conductivity and elastic constants of PEDOT: PSS with high electrical conductivity. *Macromolecules* **2015**, *48*, 585–591. [CrossRef]
19. Blackburn, J.L.; Ferguson, A.J.; Cho, C.; Grunlan, J.C. Carbon-Nanotube-Based Thermoelectric Materials and Devices. *Adv. Mater.* **2018**, *30*, 1704386. [CrossRef] [PubMed]
20. Du, Y.; Xu, J.; Paul, B.; Eklund, P. Flexible thermoelectric materials and devices. *Appl. Mater. Today* **2018**, *12*, 366–388. [CrossRef]
21. Shirakawa, H.; Louis, E.J.; MacDiarmid, A.G.; Chiang, C.K.; Heeger, A.J. Synthesis of electrically conducting organic polymers: Halogen derivatives of polyacetylene, (CH)$_x$. *J. Chem. Soc. Chem. Commun.* **1977**, 578–580. [CrossRef]
22. Kaur, G.; Adhikari, R.; Cass, P.; Bown, M.; Gunatillake, P. Electrically conductive polymers and composites for biomedical applications. *RSC Adv.* **2015**, *5*, 37553–37567. [CrossRef]
23. Pan, L.; Qiu, H.; Dou, C.; Li, Y.; Pu, L.; Xu, J.; Shi, Y. Conducting polymer nanostructures: Template synthesis and applications in energy storage. *Int. J. Mol. Sci.* **2010**, *11*, 2636–2657. [CrossRef] [PubMed]
24. Nguyen, D.N.; Yoon, H. Recent advances in nanostructured conducting polymers: From synthesis to practical applications. *Polymers* **2016**, *8*, 118. [CrossRef]
25. Hiroshige, Y.; Ookawa, M.; Toshima, N. High thermoelectric performance of poly(2,5-dimethoxyphenylenevinylene) and its derivatives. *Synth. Metals* **2006**, *156*, 1341–1347. [CrossRef]
26. Yan, H.; Toshima, N. Thermoelectric properties of alternatively layered films of polyaniline and (±)-10-camphorsulfonic acid-doped polyaniline. *Chem. Lett.* **1999**, *28*, 1217–1218. [CrossRef]
27. Yan, H.; Ohta, T.; Toshima, N. Stretched Polyaniline Films Doped by (±)-10-Camphorsulfonic Acid: Anisotropy and Improvement of Thermoelectric Properties. *Macromol. Mater. Eng.* **2001**, *286*, 139–142. [CrossRef]
28. Yao, Q.; Wang, Q.; Wang, L.; Wang, Y.; Sun, J.; Zeng, H.; Jin, Z.; Huang, X.; Chen, L. The synergic regulation of conductivity and Seebeck coefficient in pure polyaniline by chemically changing the ordered degree of molecular chains. *J. Mater. Chem. A* **2014**, *2*, 2634–2640. [CrossRef]
29. Li, J.; Tang, X.; Li, H.; Yan, Y.; Zhang, Q. Synthesis and thermoelectric properties of hydrochloric acid-doped polyaniline. *Synth. Metals* **2010**, *160*, 1153–1158. [CrossRef]
30. Kim, G.-H.; Shao, L.; Zhang, K.; Pipe, K.P. Engineered doping of organic semiconductors for enhanced thermoelectric efficiency. *Nat. Mater.* **2013**, *12*, 719. [CrossRef] [PubMed]
31. Mengistie, D.A.; Chen, C.-H.; Boopathi, K.M.; Pranoto, F.W.; Li, L.-J.; Chu, C.-W. Enhanced thermoelectric performance of PEDOT: PSS flexible bulky papers by treatment with secondary dopants. *ACS Appl. Mater. Interfaces* **2014**, *7*, 94–100. [CrossRef] [PubMed]

32. Lee, S.H.; Park, H.; Kim, S.; Son, W.; Cheong, I.W.; Kim, J.H. Transparent and flexible organic semiconductor nanofilms with enhanced thermoelectric efficiency. *J. Mater. Chem. A* **2014**, *2*, 7288–7294. [CrossRef]
33. Wang, X.; Kyaw, A.K.K.; Yin, C.; Wang, F.; Zhu, Q.; Tang, T.; Yee, P.I.; Xu, J. Enhancement of thermoelectric performance of PEDOT: PSS films by post-treatment with a superacid. *RSC Adv.* **2018**, *8*, 18334–18340. [CrossRef]
34. Zhu, H.; Liu, C.; Song, H.; Xu, J.; Kong, F.; Wang, J. Thermoelectric performance of poly(3-hexylthiophene) films doped by iodine vapor with promising high seebeck coefficient. *Electron. Mater. Lett.* **2014**, *10*, 427–431. [CrossRef]
35. Qu, S.; Yao, Q.; Wang, L.; Chen, Z.; Xu, K.; Zeng, H.; Shi, W.; Zhang, T.; Uher, C.; Chen, L. Highly anisotropic P3HT films with enhanced thermoelectric performance via organic small molecule epitaxy. *NPG Asia Mater.* **2016**, *8*, e292. [CrossRef]
36. Shi, H.; Liu, C.; Xu, J.; Song, H.; Lu, B.; Jiang, F.; Zhou, W.; Zhang, G.; Jiang, Q. Facile fabrication of PEDOT: PSS/polythiophenes bilayered nanofilms on pure organic electrodes and their thermoelectric performance. *ACS Appl. Mater. Interfaces* **2013**, *5*, 12811–12819. [CrossRef] [PubMed]
37. Lin-Chung, P.; Reinecke, T. Thermoelectric figure of merit of composite superlattice systems. *Phys. Rev. B* **1995**, *51*, 13244. [CrossRef]
38. Wang, L.; Wang, D.; Zhu, G.; Li, J.; Pan, F. Thermoelectric properties of conducting polyaniline/graphite composites. *Mater. Lett.* **2011**, *65*, 1086–1088. [CrossRef]
39. Du, Y.; Shen, S.Z.; Yang, W.; Donelson, R.; Cai, K.; Casey, P.S. Simultaneous increase in conductivity and Seebeck coefficient in a polyaniline/graphene nanosheets thermoelectric nanocomposite. *Synth. Metals* **2012**, *161*, 2688–2692. [CrossRef]
40. Wang, L.; Yao, Q.; Bi, H.; Huang, F.; Wang, Q.; Chen, L. Large thermoelectric power factor in polyaniline/graphene nanocomposite films prepared by solution-assistant dispersing method. *J. Mater. Chem. A* **2014**, *2*, 11107–11113. [CrossRef]
41. Wang, L.; Yao, Q.; Bi, H.; Huang, F.; Wang, Q.; Chen, L. PANI/graphene nanocomposite films with high thermoelectric properties by enhanced molecular ordering. *J. Mater. Chem. A* **2015**, *3*, 7086–7092. [CrossRef]
42. Zhao, Y.; Tang, G.-S.; Yu, Z.-Z.; Qi, J.-S. The effect of graphite oxide on the thermoelectric properties of polyaniline. *Carbon* **2012**, *50*, 3064–3073. [CrossRef]
43. Mitra, M.; Kulsi, C.; Chatterjee, K.; Kargupta, K.; Ganguly, S.; Banerjee, D.; Goswami, S. Reduced graphene oxide-polyaniline composites—Synthesis, characterization and optimization for thermoelectric applications. *RSC Adv.* **2015**, *5*, 31039–31048. [CrossRef]
44. Ma, W.; Liu, Y.; Yan, S.; Miao, T.; Shi, S.; Xu, Z.; Zhang, X.; Gao, C. Chemically doped macroscopic graphene fibers with significantly enhanced thermoelectric properties. *Nano Res.* **2018**, *11*, 741–750. [CrossRef]
45. Yoo, D.; Kim, J.; Kim, J.H. Direct synthesis of highly conductive poly(3,4-ethylenedioxythiophene): Poly(4-styrenesulfonate)(PEDOT: PSS)/graphene composites and their applications in energy harvesting systems. *Nano Res.* **2014**, *7*, 717–730. [CrossRef]
46. Zhang, K.; Wang, S.; Zhang, X.; Zhang, Y.; Cui, Y.; Qiu, J. Thermoelectric performance of p-type nanohybrids filled polymer composites. *Nano Energy* **2015**, *13*, 327–335. [CrossRef]
47. Du, Y.; Cai, K.; Shen, S.; Casey, P. Preparation and characterization of graphene nanosheets/poly(3-hexylthiophene) thermoelectric composite materials. *Synth. Metals* **2012**, *162*, 2102–2106. [CrossRef]
48. Wang, L.; Liu, F.; Jin, C.; Zhang, T.; Yin, Q. Preparation of polypyrrole/graphene nanosheets composites with enhanced thermoelectric properties. *RSC Adv.* **2014**, *4*, 46187–46193. [CrossRef]
49. Zhang, Z.; Chen, G.; Wang, H.; Zhai, W. Enhanced thermoelectric property by the construction of a nanocomposite 3D interconnected architecture consisting of graphene nanolayers sandwiched by polypyrrole nanowires. *J. Mater. Chem. C* **2015**, *3*, 1649–1654. [CrossRef]
50. Wang, Y.; Yang, J.; Wang, L.; Du, K.; Yin, Q.; Yin, Q. Polypyrrole/graphene/polyaniline ternary nanocomposite with high thermoelectric power factor. *ACS Appl. Mater. Interfaces* **2017**, *9*, 20124–20131. [CrossRef] [PubMed]
51. Choi, Y.; Kim, Y.; Park, S.-G.; Kim, Y.-G.; Sung, B.J.; Jang, S.-Y.; Kim, W. Effect of the carbon nanotube type on the thermoelectric properties of CNT/Nafion nanocomposites. *Org. Electron.* **2011**, *12*, 2120–2125. [CrossRef]
52. Yu, C.; Kim, Y.S.; Kim, D.; Grunlan, J.C. Thermoelectric behavior of segregated-network polymer nanocomposites. *Nano Lett.* **2008**, *8*, 4428–4432. [CrossRef] [PubMed]

53. Kim, Y.S.; Kim, D.; Martin, K.J.; Yu, C.; Grunlan, J.C. Influence of Stabilizer Concentration on Transport Behavior and Thermopower of CNT-Filled Latex-Based Composites. *Macromol. Mater. Eng.* **2010**, *295*, 431–436. [CrossRef]
54. Moriarty, G.P.; Wheeler, J.N.; Yu, C.; Grunlan, J.C. Increasing the thermoelectric power factor of polymer composites using a semiconducting stabilizer for carbon nanotubes. *Carbon* **2012**, *50*, 885–895. [CrossRef]
55. Hicks, L.; Dresselhaus, M.S. Thermoelectric figure of merit of a one-dimensional conductor. *Phys. Rev. B* **1993**, *47*, 16631. [CrossRef]
56. Meng, C.; Liu, C.; Fan, S. A promising approach to enhanced thermoelectric properties using carbon nanotube networks. *Adv. Mater.* **2010**, *22*, 535–539. [CrossRef] [PubMed]
57. Zhang, Q.; Wang, W.; Li, J.; Zhu, J.; Wang, L.; Zhu, M.; Jiang, W. Preparation and thermoelectric properties of multi-walled carbon nanotube/polyaniline hybrid nanocomposites. *J. Mater. Chem. A* **2013**, *1*, 12109–12114. [CrossRef]
58. Wang, Q.; Yao, Q.; Chang, J.; Chen, L. Enhanced thermoelectric properties of CNT/PANI composite nanofibers by highly orienting the arrangement of polymer chains. *J. Mater. Chem.* **2012**, *22*, 17612–17618. [CrossRef]
59. Liu, J.; Sun, J.; Gao, L. Flexible single-walled carbon nanotubes/polyaniline composite films and their enhanced thermoelectric properties. *Nanoscale* **2011**, *3*, 3616–3619. [CrossRef] [PubMed]
60. Yao, Q.; Chen, L.; Zhang, W.; Liufu, S.; Chen, X. Enhanced thermoelectric performance of single-walled carbon nanotubes/polyaniline hybrid nanocomposites. *Acs Nano* **2010**, *4*, 2445–2451. [CrossRef] [PubMed]
61. Wang, L.; Yao, Q.; Xiao, J.; Zeng, K.; Qu, S.; Shi, W.; Wang, Q.; Chen, L. Engineered Molecular Chain Ordering in Single-Walled Carbon Nanotubes/Polyaniline Composite Films for High-Performance Organic Thermoelectric Materials. *Chemistry* **2016**, *11*, 1804–1810. [CrossRef] [PubMed]
62. Kim, D.; Kim, Y.; Choi, K.; Grunlan, J.C.; Yu, C. Improved thermoelectric behavior of nanotube-filled polymer composites with poly(3,4-ethylenedioxythiophene) poly(styrenesulfonate). *ACS Nano* **2009**, *4*, 513–523. [CrossRef] [PubMed]
63. Yu, C.; Choi, K.; Yin, L.; Grunlan, J.C. Light-weight flexible carbon nanotube based organic composites with large thermoelectric power factors. *ACS Nano* **2011**, *5*, 7885–7892. [CrossRef] [PubMed]
64. Moriarty, G.P.; Briggs, K.; Stevens, B.; Yu, C.; Grunlan, J.C. Fully Organic Nanocomposites with High Thermoelectric Power Factors by using a Dual-Stabilizer Preparation. *Energy Technol.* **2013**, *1*, 265–272. [CrossRef]
65. Song, H.; Liu, C.; Xu, J.; Jiang, Q.; Shi, H. Fabrication of a layered nanostructure PEDOT: PSS/SWCNTs composite and its thermoelectric performance. *RSC Adv.* **2013**, *3*, 22065–22071. [CrossRef]
66. Lee, W.; Kang, Y.H.; Lee, J.Y.; Jang, K.-S.; Cho, S.Y. Improving the thermoelectric power factor of CNT/PEDOT: PSS nanocomposite films by ethylene glycol treatment. *RSC Adv.* **2016**, *6*, 53339–53344. [CrossRef]
67. Wang, L.; Jia, X.; Wang, D.; Zhu, G.; Li, J. Preparation and thermoelectric properties of polythiophene/multiwalled carbon nanotube composites. *Synth. Metals* **2013**, *181*, 79–85. [CrossRef]
68. Lee, W.; Hong, C.T.; Kwon, O.H.; Yoo, Y.; Kang, Y.H.; Lee, J.Y.; Cho, S.Y.; Jang, K.-S. Enhanced thermoelectric performance of bar-coated SWCNT/P3HT thin films. *ACS Appl. Mater. Interfaces* **2015**, *7*, 6550–6556. [CrossRef] [PubMed]
69. Bounioux, C.; Díaz-Chao, P.; Campoy-Quiles, M.; Martín-González, M.S.; Goni, A.R.; Yerushalmi-Rozen, R.; Müller, C. Thermoelectric composites of poly(3-hexylthiophene) and carbon nanotubes with a large power factor. *Energy Environ. Sci.* **2013**, *6*, 918–925. [CrossRef]
70. Hong, C.T.; Lee, W.; Kang, Y.H.; Yoo, Y.; Ryu, J.; Cho, S.Y.; Jang, K.-S. Effective doping by spin-coating and enhanced thermoelectric power factors in SWCNT/P3HT hybrid films. *J. Mater. Chem. A* **2015**, *3*, 12314–12319. [CrossRef]
71. Hong, C.T.; Kang, Y.H.; Ryu, J.; Cho, S.Y.; Jang, K.-S. Spray-printed CNT/P3HT organic thermoelectric films and power generators. *J. Mater. Chem. A* **2015**, *3*, 21428–21433. [CrossRef]
72. Wang, J.; Cai, K.; Shen, S.; Yin, J. Preparation and thermoelectric properties of multi-walled carbon nanotubes/polypyrrole composites. *Synth. Metals* **2014**, *195*, 132–136. [CrossRef]
73. Liang, L.; Chen, G.; Guo, C.-Y. Enhanced thermoelectric performance by self-assembled layered morphology of polypyrrole nanowire/single-walled carbon nanotube composites. *Compos. Sci. Technol.* **2016**, *129*, 130–136. [CrossRef]

74. Yoo, D.; Kim, J.; Lee, S.H.; Cho, W.; Choi, H.H.; Kim, F.S.; Kim, J.H. Effects of one-and two-dimensional carbon hybridization of PEDOT: PSS on the power factor of polymer thermoelectric energy conversion devices. *J. Mater. Chem. A* **2015**, *3*, 6526–6533. [CrossRef]
75. Li, X.; Liang, L.; Yang, M.; Chen, G.; Guo, C.-Y. Poly(3,4-ethylenedioxythiophene)/graphene/carbon nanotube ternary composites with improved thermoelectric performance. *Org. Electron.* **2016**, *38*, 200–204. [CrossRef]
76. Cho, C.; Stevens, B.; Hsu, J.H.; Bureau, R.; Hagen, D.A.; Regev, O.; Yu, C.; Grunlan, J.C. Completely organic multilayer thin film with thermoelectric power factor rivaling inorganic tellurides. *Adv. Mater.* **2015**, *27*, 2996–3001. [CrossRef] [PubMed]
77. Cho, C.; Wallace, K.L.; Tzeng, P.; Hsu, J.H.; Yu, C.; Grunlan, J.C. Outstanding low temperature thermoelectric power factor from completely organic thin films enabled by multidimensional conjugated nanomaterials. *Adv. Energy Mater.* **2016**, *6*, 1502168. [CrossRef]
78. Sun, Y.; Sheng, P.; Di, C.; Jiao, F.; Xu, W.; Qiu, D.; Zhu, D. Organic Thermoelectric Materials and Devices Based on p-and n-Type Poly(metal 1, 1, 2, 2-ethenetetrathiolate)s. *Adv. Mater.* **2012**, *24*, 932–937. [CrossRef] [PubMed]
79. Fukumaru, T.; Fujigaya, T.; Nakashima, N. Development of n-type cobaltocene-encapsulated carbon nanotubes with remarkable thermoelectric property. *Sci. Rep.* **2015**, *5*, 7951. [CrossRef] [PubMed]
80. Yu, C.; Murali, A.; Choi, K.; Ryu, Y. Air-stable fabric thermoelectric modules made of N-and P-type carbon nanotubes. *Energy Environ. Sci.* **2012**, *5*, 9481–9486. [CrossRef]
81. Cho, C.; Culebras, M.; Wallace, K.L.; Song, Y.; Holder, K.; Hsu, J.-H.; Yu, C.; Grunlan, J.C. Stable n-type thermoelectric multilayer thin films with high power factor from carbonaceous nanofillers. *Nano Energy* **2016**, *28*, 426–432. [CrossRef]
82. Luo, J.; Cerretti, G.; Krause, B.; Zhang, L.; Otto, T.; Jenschke, W.; Ullrich, M.; Tremel, W.; Voit, B.; Pötschke, P. Polypropylene-based melt mixed composites with singlewalled carbon nanotubes for thermoelectric applications: Switching from p-type to n-type by the addition of polyethylene glycol. *Polymer* **2017**, *108*, 513–520. [CrossRef]
83. An, C.J.; Kang, Y.H.; Song, H.; Jeong, Y.; Cho, S.Y. High-performance flexible thermoelectric generator by control of electronic structure of directly spun carbon nanotube webs with various molecular dopants. *J. Mater. Chem. A* **2017**, *5*, 15631–15639. [CrossRef]
84. Yao, Q.; Chen, L.; Xu, X.; Wang, C. The high thermoelectric properties of conducting polyaniline with special submicron-fibre structure. *Chem. Lett.* **2005**, *34*, 522–523. [CrossRef]
85. Cao, Y.; Smith, P. Liquid-crystalline solutions of electrically conducting polyaniline. *Polymer* **1993**, *34*, 3139–3143. [CrossRef]
86. De Albuquerque, J.; Melo, W.; Faria, R. Determination of physical parameters of conducting polymers by photothermal spectroscopies. *Rev. Sci. Instrum.* **2003**, *74*, 306–308. [CrossRef]
87. Möller, S.; Perlov, C.; Jackson, W.; Taussig, C.; Forrest, S.R. A polymer/semiconductor write-once read-many-times memory. *Nature* **2003**, *426*, 166. [CrossRef] [PubMed]
88. Bubnova, O.; Khan, Z.U.; Malti, A.; Braun, S.; Fahlman, M.; Berggren, M.; Crispin, X. Optimization of the thermoelectric figure of merit in the conducting polymer poly(3,4-ethylenedioxythiophene). *Nat. Mater.* **2011**, *10*, 429. [CrossRef] [PubMed]
89. Park, T.; Park, C.; Kim, B.; Shin, H.; Kim, E. Flexible PEDOT electrodes with large thermoelectric power factors to generate electricity by the touch of fingertips. *Energy Environ. Sci.* **2013**, *6*, 788–792. [CrossRef]
90. Scholdt, M.; Do, H.; Lang, J.; Gall, A.; Colsmann, A.; Lemmer, U.; Koenig, J.D.; Winkler, M.; Boettner, H. Organic semiconductors for thermoelectric applications. *J. Electr. Mater.* **2010**, *39*, 1589–1592. [CrossRef]
91. Liu, C.; Lu, B.; Yan, J.; Xu, J.; Yue, R.; Zhu, Z.; Zhou, S.; Hu, X.; Zhang, Z.; Chen, P. Highly conducting free-standing poly(3,4-ethylenedioxythiophene)/poly (styrenesulfonate) films with improved thermoelectric performances. *Synth. Metals* **2010**, *160*, 2481–2485. [CrossRef]
92. Zhang, K.; Qiu, J.; Wang, S. Thermoelectric properties of PEDOT nanowire/PEDOT hybrids. *Nanoscale* **2016**, *8*, 8033–8041. [CrossRef] [PubMed]
93. Ouyang, J. "Secondary doping" methods to significantly enhance the conductivity of PEDOT: PSS for its application as transparent electrode of optoelectronic devices. *Displays* **2013**, *34*, 423–436. [CrossRef]
94. Kim, J.; Jung, J.; Lee, D.; Joo, J. Enhancement of electrical conductivity of poly(3,4-ethylenedioxythiophene)/poly(4-styrenesulfonate) by a change of solvents. *Synth. Metals* **2002**, *126*, 311–316. [CrossRef]

95. Culebras, M.; Gómez, C.; Cantarero, A. Enhanced thermoelectric performance of PEDOT with different counter-ions optimized by chemical reduction. *J. Mater. Chem. A* **2014**, *2*, 10109–10115. [CrossRef]
96. Wang, G.; Huang, W.; Eastham, N.D.; Fabiano, S.; Manley, E.F.; Zengg, L.; Wang, B.; Zhang, X.; Chend, Z.; Lib, R.; et al. Aggregation control in natural brush-printed conjugated polymer films and implications for enhancing charge transport. *Proc. Natl. Acad. Sci. USA* **2017**, *114*, E10066. [CrossRef] [PubMed]
97. Zhu, Z.; Liu, C.; Shi, H.; Jiang, Q.; Xu, J.; Jiang, F.; Xiong, J.; Liu, E. An effective approach to enhanced thermoelectric properties of PEDOT: PSS films by a DES post-treatment. *J. Polym. Sci. Part B Polym. Phys.* **2015**, *53*, 885–892. [CrossRef]
98. Zhang, Q.; Vigier, K.D.O.; Royer, S.; Jerome, F. Deep eutectic solvents: Syntheses, properties and applications. *Chem. Soc. Rev.* **2012**, *41*, 7108–7146. [CrossRef] [PubMed]
99. Kline, R.J.; McGehee, M.D.; Kadnikova, E.N.; Liu, J.; Fréchet, J.M.; Toney, M.F. Dependence of regioregular poly (3-hexylthiophene) film morphology and field-effect mobility on molecular weight. *Macromolecules* **2005**, *38*, 3312–3319. [CrossRef]
100. Teehan, S.; Efstathiadis, H.; Haldar, P. Thermoelectric power factor enhancement of AZO/In-AZO quantum well multilayer structures as compared to bulk films. *J. Alloys Compounds* **2012**, *539*, 129–136. [CrossRef]
101. Lee, H.J.; Anoop, G.; Lee, H.J.; Kim, C.; Park, J.-W.; Choi, J.; Kim, H.; Kim, Y.-J.; Lee, E.; Lee, S.-G. Enhanced thermoelectric performance of PEDOT: PSS/PANI–CSA polymer multilayer structures. *Energy Environ. Sci.* **2016**, *9*, 2806–2811. [CrossRef]
102. McGrail, B.T.; Sehirlioglu, A.; Pentzer, E. Polymer composites for thermoelectric applications. *Angew. Chem. Int. Ed.* **2015**, *54*, 1710–1723. [CrossRef] [PubMed]
103. Dey, A.; Bajpai, O.P.; Sikder, A.K.; Chattopadhyay, S.; Khan, M.A.S. Recent advances in CNT/graphene based thermoelectric polymer nanocomposite: A proficient move towards waste energy harvesting. *Renew. Sustain. Energy Rev.* **2016**, *53*, 653–671. [CrossRef]
104. Xiang, J.; Drzal, L.T. Templated growth of polyaniline on exfoliated graphene nanoplatelets (GNP) and its thermoelectric properties. *Polymer* **2012**, *53*, 4202–4210. [CrossRef]
105. Chatterjee, K.; Mitra, M.; Kargupta, K.; Ganguly, S.; Banerjee, D. Synthesis, characterization and enhanced thermoelectric performance of structurally ordered cable-like novel polyaniline–bismuth telluride nanocomposite. *Nanotechnology* **2013**, *24*, 215703. [CrossRef] [PubMed]
106. Kim, G.H.; Hwang, D.H.; Woo, S.I. Thermoelectric properties of nanocomposite thin films prepared with poly(3, 4-ethylenedioxythiophene) poly(styrenesulfonate) and graphene. *Phys. Chem. Chem. Phys.* **2012**, *14*, 3530–3536. [CrossRef] [PubMed]
107. Vavro, J.; Llaguno, M.C.; Satishkumar, B.; Luzzi, D.E.; Fischer, J.E. Electrical and thermal properties of C 60-filled single-wall carbon nanotubes. *Appl. Phys. Lett.* **2002**, *80*, 1450–1452. [CrossRef]
108. Zhang, K.; Zhang, Y.; Wang, S. Enhancing thermoelectric properties of organic composites through hierarchical nanostructures. *Sci. Rep.* **2013**, *3*, 3448. [CrossRef] [PubMed]
109. Paloheimo, J.; Isotalo, H.; Kastner, J.; Kuzmany, H. Conduction mechanisms in undoped thin films of C60 and C60/70. *Synth. Metals* **1993**, *56*, 3185–3190. [CrossRef]
110. Lu, Y.; Song, Y.; Wang, F. Thermoelectric properties of graphene nanosheets-modified polyaniline hybrid nanocomposites by an in situ chemical polymerization. *Mater. Chem. Phys.* **2013**, *138*, 238–244. [CrossRef]
111. Han, S.; Zhai, W.; Chen, G.; Wang, X. Morphology and thermoelectric properties of graphene nanosheets enwrapped with polypyrrole. *RSC Adv.* **2014**, *4*, 29281–29285. [CrossRef]
112. Dresselhaus, M.; Dresselhaus, G.; Hofmann, M. Other one-dimensional systems and thermal properties. *J. Vacuum Sci. Technol. B Microelectron. Nanometer Struct. Proc. Meas. Phenom.* **2008**, *26*, 1613–1618. [CrossRef]
113. Botiz, I.; Stingelin, N. Influence of molecular conformations and microstructure on the optoelectronic properties of conjugated polymers. *Materials* **2014**, *7*, 2273–2300. [CrossRef] [PubMed]
114. See, K.C.; Feser, J.P.; Chen, C.E.; Majumdar, A.; Urban, J.J.; Segalman, R.A. Water-processable polymer–nanocrystal hybrids for thermoelectrics. *Nano Lett.* **2010**, *10*, 4664–4667. [CrossRef] [PubMed]
115. Zhang, Z.; Chen, G.; Wang, H.; Li, X. Template-Directed In Situ Polymerization Preparation of Nanocomposites of PEDOT: PSS-Coated Multi-Walled Carbon Nanotubes with Enhanced Thermoelectric Property. *Chemistry* **2015**, *10*, 149–153.
116. Lin, Y.-J.; Ni, W.-S.; Lee, J.-Y. Effect of incorporation of ethylene glycol into PEDOT: PSS on electron phonon coupling and conductivity. *J. Appl. Phys.* **2015**, *117*, 215501.

117. Rivadulla, F.; Mateo-Mateo, C.; Correa-Duarte, M. Layer-by-layer polymer coating of carbon nanotubes: Tuning of electrical conductivity in random networks. *J. Am. Chem. Soc.* **2010**, *132*, 3751–3755. [CrossRef] [PubMed]
118. Liu, J.; Qiu, L.; Portale, G.; Koopmans, M.; Ten Brink, G.; Hummelen, J.C.; Koster, L.J.A. N-Type Organic Thermoelectrics: Improved Power Factor by Tailoring Host–Dopant Miscibility. *Adv. Mater.* **2017**, *29*, 1701641. [CrossRef] [PubMed]
119. Hwang, S.; Potscavage, W.J.; Yang, Y.S.; Park, I.S.; Matsushima, T.; Adachi, C. Solution-processed organic thermoelectric materials exhibiting doping-concentration-dependent polarity. *Phys. Chem. Chem. Phys.* **2016**, *18*, 29199–29207. [CrossRef] [PubMed]
120. Collins, P.G.; Bradley, K.; Ishigami, M.; Zettl, D.A. Extreme oxygen sensitivity of electronic properties of carbon nanotubes. *Science* **2000**, *287*, 1801–1804. [CrossRef] [PubMed]
121. Wan, C.; Gu, X.; Dang, F.; Itoh, T.; Wang, Y.; Sasaki, H.; Kondo, M.; Koga, K.; Yabuki, K.; Snyder, G.J. Flexible n-type thermoelectric materials by organic intercalation of layered transition metal dichalcogenide TiS_2. *Nat. Mater.* **2015**, *14*, 622. [CrossRef] [PubMed]
122. Watts, P.C.; Mureau, N.; Tang, Z.; Miyajima, Y.; Carey, J.D.; Silva, S.R.P. The importance of oxygen-containing defects on carbon nanotubes for the detection of polar and non-polar vapours through hydrogen bond formation. *Nanotechnology* **2007**, *18*, 175701. [CrossRef]
123. Yoo, D.; Lee, J.J.; Park, C.; Choi, H.H.; Kim, J.-H. N-type organic thermoelectric materials based on polyaniline doped with the aprotic ionic liquid 1-ethyl-3-methylimidazolium ethyl sulfate. *RSC Adv.* **2016**, *6*, 37130–37135. [CrossRef]
124. Zuo, G.; Li, Z.; Wang, E.; Kemerink, M. High Seebeck Coefficient and Power Factor in n-Type Organic Thermoelectrics. *Adv. Electron. Mater.* **2018**, *4*, 1700501. [CrossRef]
125. Hewitt, C.A.; Kaiser, A.B.; Roth, S.; Craps, M.; Czerw, R.; Carroll, D.L. Multilayered carbon nanotube/polymer composite based thermoelectric fabrics. *Nano Lett.* **2012**, *12*, 1307–1310. [CrossRef] [PubMed]
126. Nonoguchi, Y.; Nakano, M.; Murayama, T.; Hagino, H.; Hama, S.; Miyazaki, K.; Matsubara, R.; Nakamura, M.; Kawai, T. Simple Salt-Coordinated n-Type Nanocarbon Materials Stable in Air. *Adv. Funct. Mater.* **2016**, *26*, 3021–3028. [CrossRef]
127. Ryu, Y.; Freeman, D.; Yu, C. High electrical conductivity and n-type thermopower from double-/single-wall carbon nanotubes by manipulating charge interactions between nanotubes and organic/inorganic nanomaterials. *Carbon* **2011**, *49*, 4745–4751. [CrossRef]
128. Freeman, D.D.; Choi, K.; Yu, C. N-type thermoelectric performance of functionalized carbon nanotube-filled polymer composites. *PLoS ONE* **2012**, *7*, e47822. [CrossRef] [PubMed]
129. Snyder, G.J.; Toberer, E.S. Complex thermoelectric materials. In *Materials For Sustainable Energy: A Collection of Peer-Reviewed Research and Review Articles from Nature Publishing Group*; World Scientific: Singapore, 2011; pp. 101–110.
130. Zhao, L.-D.; Lo, S.-H.; Zhang, Y.; Sun, H.; Tan, G.; Uher, C.; Wolverton, C.; Dravid, V.P.; Kanatzidis, M.G. Ultralow thermal conductivity and high thermoelectric figure of merit in SnSe crystals. *Nature* **2014**, *508*, 373. [CrossRef] [PubMed]
131. Kishore, R.; Priya, S. A Review on Low-Grade Thermal Energy Harvesting: Materials, Methods and Devices. *Materials* **2018**, *11*, 1433. [CrossRef] [PubMed]
132. Wei, Q.; Mukaida, M.; Kirihara, K.; Naitoh, Y.; Ishida, T. Polymer thermoelectric modules screen-printed on paper. *RSC Adv.* **2014**, *4*, 28802–28806. [CrossRef]
133. Dörling, B.; Ryan, J.D.; Craddock, J.D.; Sorrentino, A.; Basaty, A.E.; Gomez, A.; Garriga, M.; Pereiro, E.; Anthony, J.E.; Weisenberger, M.C. Photoinduced p-to n-type Switching in Thermoelectric Polymer-Carbon Nanotube Composites. *Adv. Mater.* **2016**, *28*, 2782–2789. [CrossRef] [PubMed]
134. Wu, G.; Gao, C.; Chen, G.; Wang, X.; Wang, H. High-performance organic thermoelectric modules based on flexible films of a novel n-type single-walled carbon nanotube. *J. Mater. Chem. A* **2016**, *4*, 14187–14193. [CrossRef]
135. An, C.J.; Kang, Y.H.; Lee, A.-Y.; Jang, K.-S.; Jeong, Y.; Cho, S.Y. Foldable thermoelectric materials: Improvement of the thermoelectric performance of directly spun CNT webs by individual control of electrical and thermal conductivity. *ACS Appl. Mater. Interfaces* **2016**, *8*, 22142–22150. [CrossRef] [PubMed]

136. Culebras, M.; Cho, C.; Krecker, M.; Smith, R.; Song, Y.; Gómez, C.M.; Cantarero, A.S.; Grunlan, J.C. High thermoelectric power factor organic thin films through combination of nanotube multilayer assembly and electrochemical polymerization. *ACS Appl. Mater. Interfaces* **2017**, *9*, 6306–6313. [CrossRef] [PubMed]
137. Wu, Q.; Hu, J. Waterborne polyurethane based thermoelectric composites and their application potential in wearable thermoelectric textiles. *Compo. Part B Eng.* **2016**, *107*, 59–66. [CrossRef]
138. Du, Y.; Cai, K.; Chen, S.; Wang, H.; Shen, S.Z.; Donelson, R.; Lin, T. Thermoelectric fabrics: Toward power generating clothing. *Sci. Rep.* **2015**, *5*, 6411. [CrossRef] [PubMed]
139. Ryan, J.D.; Lund, A.; Hofmann, A.I.; Kroon, R.; Sarabia-Riquelme, R.; Weisenberger, M.C.; Müller, C. All-Organic Textile Thermoelectrics with Carbon-Nanotube-Coated n-Type Yarns. *ACS Appl. Energy Mater.* **2018**. [CrossRef] [PubMed]

© 2018 by the authors. Licensee MDPI, Basel, Switzerland. This article is an open access article distributed under the terms and conditions of the Creative Commons Attribution (CC BY) license (http://creativecommons.org/licenses/by/4.0/).

Article

Mesoporous Highly-Deformable Composite Polymer for a Gapless Triboelectric Nanogenerator via a One-Step Metal Oxidation Process

Hee Jae Hwang [1], Younghoon Lee [1], Choongyeop Lee [1], Youngsuk Nam [1], Jinhyoung Park [2], Dukhyun Choi [1,*] and Dongseob Kim [3,*]

1. Department of Mechanical Engineering, Kyung Hee University, 1732 Deogyeong-daero, Giheung-gu, Yongin-Si, Gyeonggi-do 446701, Korea; hjhwang@khu.ac.kr (H.J.H.); younghoon@snu.ac.kr (Y.L.); cylee@khu.ac.kr (C.L.); ysnam1@khu.ac.kr (Y.N.)
2. Construction Equipment R&D Group, Korea Institute of Industrial Technology (KITECH), 288-1, Daehak-ri, Hayang-eup, Gyeongsan-si, Gyeongsangbuk-do 712091, Korea; jh.park@kitech.re.kr
3. Aircraft System Technology Group, Korea Institute of Industrial Technology (KITECH), 57, Yangho-gil, Yeongcheon-si, Gyeongbuk-do 38822, Korea
* Correspondence: dchoi@khu.ac.kr (D.C.); yusae@kitech.re.kr (D.K.); Tel.: +82-31-201-3320 (D.C.); +82-54-339-0636 (D.K.)

Received: 16 November 2018; Accepted: 8 December 2018; Published: 11 December 2018

Abstract: The oxidation of metal microparticles (MPs) in a polymer film yields a mesoporous highly-deformable composite polymer for enhancing performance and creating a gapless structure of triboelectric nanogenerators (TENGs). This is a one-step scalable synthesis for developing large-scale, cost-effective, and light-weight mesoporous polymer composites. We demonstrate mesoporous aluminum oxide (Al_2O_3) polydimethylsiloxane (PDMS) composites with a nano-flake structure on the surface of Al_2O_3 MPs in pores. The porosity of mesoporous Al_2O_3-PDMS films reaches 71.35% as the concentration of Al MPs increases to 15%. As a result, the film capacitance is enhanced 1.8 times, and TENG output performance is 6.67-times greater at 33.3 kPa and 4 Hz. The pressure sensitivity of 6.71 V/kPa and 0.18 µA/kPa is determined under the pressure range of 5.5–33.3 kPa. Based on these structures, we apply mesoporous Al_2O_3-PDMS film to a gapless TENG structure and obtain a linear pressure sensitivity of 1.00 V/kPa and 0.02 µA/kPa, respectively. Finally, we demonstrate self-powered safety cushion sensors for monitoring human sitting position by using gapless TENGs, which are developed with a large-scale and highly-deformable mesoporous Al_2O_3-PDMS film with dimensions of 6 × 5 pixels (33 × 27 cm^2).

Keywords: mesoporous composite polymer; metal oxidation; gapless; triboelectric nanogenerator; high deformability

---

1. Introduction

Mechanical energy is the most common energy in our surroundings, and it can be converted into electricity anytime and anywhere. Triboelectric nanogenerators (TENGs), which are power-generating devices introduced in 2012, have been proven as cost effective, simple, able to cover large areas, durable, and efficient for energy harvesting [1–3]. Due to their simple mechanism, which utilizes repeating cycles of contact and separation between two materials, TENGs have demonstrated promising capability in scavenging energy and as variable sensors of mechanical energy (vibration, rotating, etc.), ocean waves, fluids, and human activities. Further, TENGs can generate electric signals without relying on external power sources, thus enabling the development of self-powered Internet of Things (IoT) sensors [4,5].

With a contact-separation structure, the surface charge density of the tribo-material is the key to achieving a high TENG output performance. To increase the surface charge, many fabrications and systems have been introduced in previous studies such as micro-patterned arrays, porous structures, multilayer alignment, ion injections, ground systems, mixing of high dielectric constant materials, and charging pumps [6–12]. Among them, it is still requested to develop a new porous fabrication technique to overcome the limits of fabricating porous structures. The reasons for studying porous structures is not only limited to increasing the output performance, but also applies to life because foam materials apply to many fields such as matrix structures, seat cushions, and shoe inserts. In previous studies, the fabrication of porous structures has generally consisted of two steps, suspensions and removal, which include techniques such as solvent casting/particle leaching and melting molds. However, they have more two steps in their fabrication methods and require more time to fabricate than TENG samples. Furthermore, these fabrications are hard to scale up and suffer from a lack of control over end product porosity [7,13–15].

In this work, we report a facile one-step metal oxidation process for creating mesoporous polymer films with a highly deformable composite structure to improve the capacitance significantly, thus resulting in enhancing the output performance of TENGs. The highly-deformable composite polymers are fabricated by casting a mixture of polymer mixed metal microparticle (MP) solution, like polydimethylsiloxane (PDMS), deionized (DI) water, and aluminum (Al) MPs, followed by evaporating water in an 80 °C oven, where the concentrations of Al MPs are in the range of 0–15 wt%. Interestingly, it is found that the surface of Al MPs is oxidized, and nano-flake structures are formed. As a result, we can obtain a mesoporous PDMS polymer integrated with aluminum oxide (Al_2O_3) MPs with the nano-flake surfaces. Based on the mesoporous highly-deformable PDMS composite polymer, the TENG output shows significant enhancement, as compared to a non-porous PDMS. The mesoporous PDMS composite with the weight fraction of 15 wt% exhibits the highest voltage of 259.58 V and current of 7.16 μA under 33.3 kPa and at 4 Hz. For a practical application, we demonstrate a self-powered safety cushion sensor with optimized TENG output performance using a low-cost, simple synthesis and that is light-weight to cover a large area of 33×27 cm^2 (6×5 pixels).

2. Materials and Methods

2.1. The Fabrication Process of the Mesoporous Al_2O_3-PDMS

A mesoporous Al_2O_3-PDMS structure was fabricated by a one-step method, as shown in Figure 1a. In a previous study, porous PDMS embedded powder was made by a two-step mixing and etching process. However, in the present study, we introduce a one-step method for the porous structure. The PDMS was prepared as a mixture of base resin and curing agent (Sylgard 184 A:Sylgard 184 B, Dow Corning Co., Midland, MI, USA) at a weight ratio of 10:1. Al MPs (1 μm–10 μm), and DI water was added to the solution at weight ratios from 5%–15% and 50%. After degassing under vacuum for 20 min, the solution was laminated on the Teflon mold at a thickness of 3 mm and cured at 80 °C for six hours. The mesoporous Al_2O_3-PDMS was then peeled off the Teflon mold. Al tape (70 μm) was attached to the bottom of mesoporous Al_2O_3-PDMS by the electrode layer.

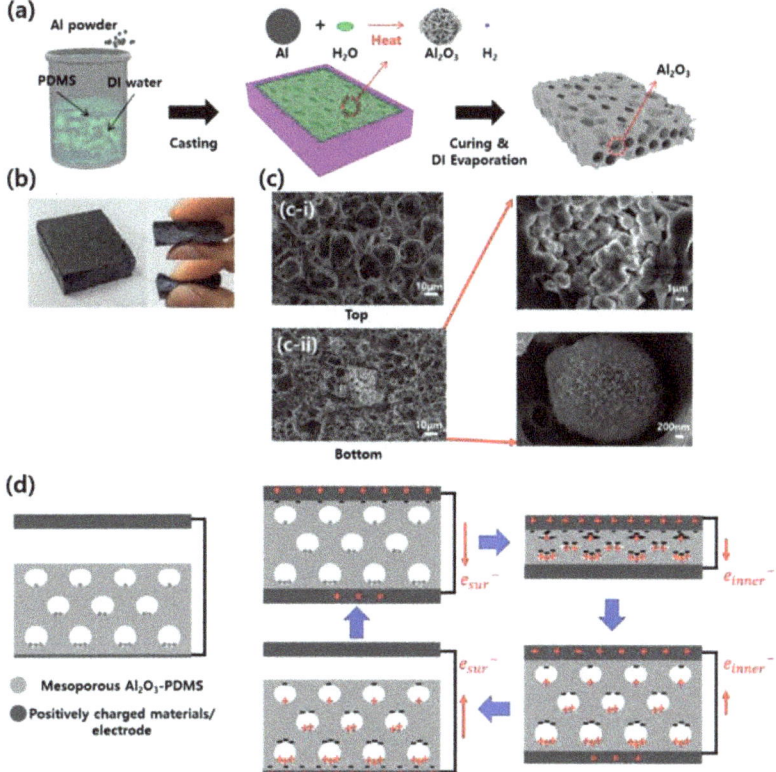

Figure 1. Mesoporous highly-deformable polymer composite and triboelectric nanogenerator (TENG) application. (**a**) Schematic diagram of mesoporous Al_2O_3-PDMS fabrication process. (**b**) Photos of highly-deformable mesoporous Al_2O_3-PDMS composite polymer. (**c**) The scanning electron microscopy (SEM) images of mesoporous Al_2O_3-PDMS top (**c-i**) and bottom (**c-ii**) parts. (**d**) Mechanism of the electricity generation from the mesoporous Al_2O_3-PDMS TENG under external force.

2.2. Fabrication and Output Measurement of the TENGs

To design the vertical contact-separation mode TENG, the top electrode was prepared with a 3 × 3 cm² 30-µm-thick conductive graphene tape (DASAN Solueta, SSC30) and mesoporous Al_2O_3-PDMS, with a thickness of 3 mm. The conductive graphene tape and mesoporous Al_2O_3-PDMS were attached to the acryl with double-sided tape.

To apply the vertical contact-separation mode, we used a pushing tester (JPT-110) ranging from 5 N–30 N of force. To measure TENG output performance, we used an oscilloscope (Tektronix MDO 3052, Beaverton, OR, USA) probe with 40 MΩ and a preamplifier (Stanford Research Systems, SR570, Sunnyvale, CA, USA) connected to the MDO 3052. TENG performance was measured at a working frequency of 4 Hz, and the gap distance was 4 mm between the two materials.

2.3. Analysis of the Al_2O_3-PDMS TENG

The surface morphology, cross-section, energy dispersive spectroscopic (EDS) mapping, and line scan characteristics of the PDMS layer and mesoporous Al_2O_3-PDMS were examined using high-resolution field emitting scanning electron microscopy (HR FE-SEM, Leo supra 55, Carl Zeiss, Stockholm, Sweden). The porosity was measured by changes of volume and mass. Compressive strength was tested using a Universal Testing System (Instron, MA, USA) following ASTM D

695. Capacitance measurements of the mesoporous Al_2O_3-PDMS films were performed at room temperature, at an electrical frequency of 2000 Hz and a bias of 100 mV using an E4980a precision inductance, capacitance, and resistance meter (Agilent Technologies, Santa Clara, CA, USA). A custom-built sensor probe station with a programmable x-, y-, and z-axis stage (0.1-μm resolution) enabled the mesoporous Al_2O_3-PDMS films to be used to obtain the exact pressure, and a force gauge (Mark-10, Copiague, NY, USA) measured the load.

2.4. Fabrication of the Self-Powered Safety Cushion Sensors

We prepared 30 ea, to make a 6 × 5 pixels, of mesoporous Al_2O_3-PDMS films attached by conductive graphene tape with a size of 3 × 3 cm^2 for both the top and bottom side. We put them on the bottom polyethylene terephthalate (PET) film, and the total size was 33 × 27 cm^2, 6 × 5 pixels, with 3 cm of distance between each cell, then we put the PET film on them. Each cell was connected by a wire to the DAQ 6011, multichannel. It could show the position pressure by color: green was assigned to 1.3–2.5 V, yellow from 2.5–3.8 V, orange from 3.8–5 V, and red over 5 V in real time.

3. Results and Discussion

The schematic diagrams for the synthesis process of the mesoporous Al_2O_3-PDMS-based TENGs are shown in Figure 1a. In previous studies, researchers generally made porous structure by using NaCl, ZnO, or polystyrene particles in the PDMS and removing them by immersion in water solution or by chemical treatments. In this study, we used a one-step method without removing anything. The advantages of a one-step method were two-fold. One was a chemical reaction increasing the number and size of pores, given as:

$$2Al + 3H_2O \xrightarrow{heat} Al_2O_3 + 3H_2. \qquad (1)$$

Al can easily react with H_2O because it is an active metal like K, Ca, Na etc., at room temperature or under the drying conditions (80 °C). H_2 gas, one product of the chemical reaction, increased the volume by increasing the size and the number of pores. Furthermore, Al_2O_3 particles had a nano-flake surface following the reaction, increasing the surface area to increase the output performance, as shown in Figure 1c-ii. By most aspects (costs, characteristics like the output performance, particle size), we could choose Al as a suitable active metal and tribo-material. It also has higher density than PDMS and DI water. In a previous study, to increase the output performance, the authors had constructed a multi-layered structure, alternating positively- and negatively-charged layers. In this study, Al, positive material, sank to the bottom because of its relatively higher density than PDMS and DI water. As a result, the top layer was more negatively charged than the bottom layer [8]. The resultant mesoporous Al_2O_3-PDMS had a multi-layered structure, and this may have increased the output performance. A photograph of the mesoporous Al_2O_3-PDMS (30 × 30 × 7 mm^3) and the scanning electron microscope image are shown in Figure 1b,c. More Al_2O_3 is seen at the bottom (Figure 1c-ii) than at the top (Figure 1c-i).

Figure 1d shows schematically the structure of the mesoporous Al_2O_3-PDMS device consisting of a top electrode, mesoporous Al_2O_3-PDMS, and the bottom electrode under the vertical contact-separation mode. When the two materials make contact, the top electrode is positively charged and mesoporous Al_2O_3-PDMS is negatively charged. In this process, we can separate the two steps, contact and compression. At first, in the contact process, the top electrode firstly contacts the surface of mesoporous Al_2O_3-PDMS; it is positively charged, and electrons ($e^-{}_{sur}$) are moved to the bottom electrode. Then, in the compression process, because of the porous structure of mesoporous Al_2O_3-PDMS, the PDMS in the inner surface of pore contacts Al_2O_3, and positive (micro-flake Al_2O_3) and negative (PDMS) charges increase, additional electrons ($e^-{}_{inner}$) are moved to the bottom electrode to neutralize the positive charges and the top electrode. Overall, the negative and positive charges occur at not only the surface, but also the inner pores of mesoporous Al_2O_3-PDMS, and the output performance is increased.

When the external force is removed, firstly, the inner surface and Al_2O_3 separate, and e^-_{inner} move to the top electrode. Secondly, the top electrode and surface of mesoporous Al_2O_3-PDMS separate, and e^-_{sur} move to the top electrode to neutralize the negative charges in the bottom electrode; thus, charges occur not only at the surface of mesoporous Al_2O_3-PDMS, but also in the inner pore of mesoporous Al_2O_3-PDMS, and stacked charges increase. Stacked charges increase due to the advantage of the porous structure (double contact) and can increase the output performance of TENGs. The charge and the voltage were calculated by the following equation:

$$\sigma_P = \sigma_{sur} + \sigma_{inner} \tag{2}$$

$$V_{OC} = \frac{(\sigma_{sur} + \sigma_{inner})x(t)}{\varepsilon_0} \tag{3}$$

Here, σ_P is the total charge, σ_{sur} is the surface charge, σ_{inner} is the inner pore charge, V_{OC} is the open-circuit voltage, and ε_0 is the permittivity in a vacuum.

To confirm the additional advantages of Al_2O_3 on TENG output performance, we compared the output performance of other materials. We compared the output performance (voltage, current) of metal oxides such as Al, Al_2O_3, TiO_2, SiO_2, and HfO by sputtering them on an Al sheet (S-Al, S-Al_2O_3, S-TiO_2, S-SiO_2, S-HfO) and other materials (alkali-treated micro-nano-flaked aluminum hydroxide ($Al(OH)_3$), polished Al (P-Al), Al sheet dipped in DI water and dried (Al_2O_3)), as shown in Table 1. The results showed that sputtered Al_2O_3 had the best output performance (40.96 V, 2.86 µA). In previous studies, materials with greater dielectric constants had better output performance [16]. Although TiO_2 (ε_r = 80) and HfO (ε_r = 25) have larger dielectric constants than Al_2O_3 (ε_r = 9), they had lower output performance than Al_2O_3. Based on this result, we believe that Al_2O_3 was suitable for fabricating the porous structure and increasing the output performance.

Table 1. Comparison of TENG output performance with various metal oxides and fabrication techniques (sputtering method: S-Al_2O_3, S-Al, S-TiO_2, S-SiO_2, S-HfO/Alkali treatment: $Al(OH)_3$, dipping in water and drying: Al_2O_3, polishing: P-Al).

The Output Performance	S-Al_2O_3	Al_2O_3	S-SiO_2	S-TiO_2	S-HfO	S-Al	Al(ref)	P-Al	$Al(OH)_3$
Voltage (V)	40.97	36.3	37.30	33.65	31.42	32.29	31.57	36.32	31.42
Current (µA)	2.86	2.44	2.24	2.59	1.92	1.98	1.19	2.82	2.09

Figure 2 shows the characteristics of mesoporous Al_2O_3-PDMS. First, we confirmed that Al powder changed to Al_2O_3 according to the chemical reaction in Equation (1). By energy dispersive spectrometry, we confirmed that the elements Si, Al, C, and O were present. Figure 2a,b shows the SEM and EDS images. Through the SEM image, we confirmed the location of ball-like Al_2O_3. By comparison of the SEM images and the distribution of Al (7.48%) and O (28.28%) via EDS images, we see that Al powder, mixed with PDMS and DI water, reacted to form Al_2O_3 powder after drying (for 1 h at 80 °C). In previous studies by J. Chun et al., a mesoporous structure made by DI water and demonstrating pores impregnated with Au nanoparticles had 59% of the maximum porosity at a 50% DI water concentration. However, in this study, at the same condition, we got 45.4% of the maximum porosity. When Al MPs were embedded, the porosity increased to approximately 71.35% as the amount of Al increased to 15%, as shown in Figure 2c. The meaning of high porosity is that the mesoporous film is much lighter, more flexible, and deformable. At the same volume, we can get mass ratios of 1:0.37:0.35:0.27 of 0 wt%:5 wt%:10 wt%:15 wt% of Al MP concentration.

Figure 2d shows the deformability of mesoporous Al_2O_3-PDMS (5, 10, 15 wt%) films with a diameter size of 50 mm and a height of 13 mm by ASTM D-695. We limited the jig amplitude to 8 mm and measured the maximum compressed stress at the same conditions. When we compared them, the compressive stress of a mesoporous Al_2O_3-PDMS (5 wt%) film was 148.5 MPa with 58.3% of the porosity. However, in the case of mesoporous Al_2O_3-PDMS (15 wt%) films, because they had 71.35%

of porosity, the maximum compressed stress was the minimum value, 11.7 MPa. That means low compressed stress can be easily deformed, and it is easy to fully contact between the top electrode and a mesoporous Al$_2$O$_3$-PDMS, like the inset of Figure 2d. We believe that the high porosity and light weight are major factors; if their values increase, they can be higher deformability and can increase the relative capacitance change and output performance of TENGs.

Figure 2e,f shows that mesoporous Al$_2$O$_3$-PDMS is much more deformable than non-porous PDMS and mesoporous PDMS (0 wt%). To compare the deformability of non-porous PDMS and mesoporous Al$_2$O$_3$-PDMS (0, 5, 10, 15 wt%), we measured the relative capacitance with $7 \times 7 \times 3$ mm^2 films at 0, 16.7, and 33.3 kPa, given respectively as C_0, $C_{16.7kPa}$, and $C_{33.3kPa}$. When non-porous PDMS, 0, 5, 10, 15 wt% mesoporous Al$_2$O$_3$-PDMS, is pressed at 33.3 kPa ($C_{33.3kPa}$), the max relative capacitance was 1.08, 1.69, 1.97, 2.07, and 3.03, respectively, at 33.3 kPa. When comparing 0 wt% and 15 wt% of mesoporous Al$_2$O$_3$-PDMS, the capacitance of the film increased 1.79-times more. Generally, capacitance is expressed by the following Equation (3):

$$C = \varepsilon_0 \varepsilon_r \frac{A}{d}. \tag{4}$$

Here, C is the capacitance, ε_0 is the permittivity in a vacuum, ε_r is the relative permittivity, A is the area of the film, and d is the thickness of the film. Based on this equation, if ε_r and A are constants, the capacitance inversely depends on the thickness of the film. Thus, as the film is deformable, the film thickness can be easily compressed, and the capacitance increases.

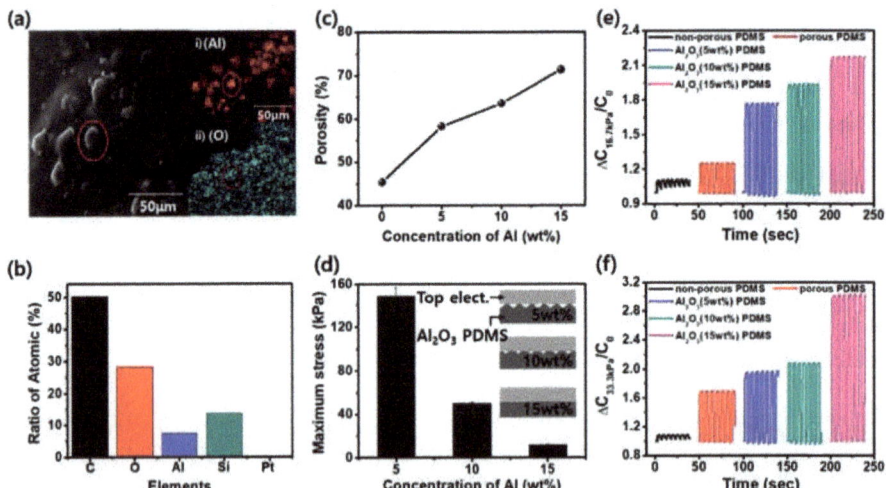

Figure 2. Characterizations of mesoporous Al$_2$O$_3$-PDMS. (**a**) SEM and EDS images of the elemental maps of (i) Al (red) and (ii) O (cyan); (**b**) ratio of atomic % of mesoporous Al$_2$O$_3$-PDMS; (**c**) porosity; (**d**) compressed strength at a deformation of 8 mm; the inset is the schematic of the intensity of deformation by compressed strength; and (**e**,**f**) relative capacitance change ($\Delta C_{16.7kPa}$, $\Delta C_{33.3kPa}$) to the initial capacitance under no pressure (C_0) of non-porous PDMS and mesoporous Al$_2$O$_3$-PDMS (0, 5, 10, 15 wt%).

Figure 3 shows the output voltage and output current of non-porous PDMS and P-Al$_2$O$_3$-PDMS (5, 10, 15 wt%). First, we confirmed the advantage of the porous structure. In Figure 3a, we compare non-porous PDMS and mesoporous P-Al$_2$O$_3$-PDMS (15 wt%) at 30 N of force, a 4-Hz frequency, a 4-mm gap distance, and a 3×3 cm^2 active area. The output voltage of non-porous PDMS and mesoporous Al$_2$O$_3$-PDMS (15 wt%) was 39 V and 259.58 V, respectively. The mesoporous structure film had 6.66-times greater output voltage than the non-porous PDMS film. To understand the influence of porous structure in mesoporous Al$_2$O$_3$-PDMS, we compared one cycle of the output voltage generated by non-porous PDMS and mesoporous Al$_2$O$_3$-PDMS. In non-porous PDMS, the contact and releasing peaks took 34 and 32 ms to generate electric energy, as shown in Figure 3a-i. However, in mesoporous Al$_2$O$_3$-PDMS, contact and releasing peaks took 70 and 77 ms, 2.22-times longer, as shown in Figure 3a-ii. The reason is structural. In a previous study, soft membranes and porous structures have been shown to increase the output performance by maximizing effective contact area and the potential difference produced between the Al$_2$O$_3$ MPs and PDMS inside pores by contact due to different triboelectric tendencies.

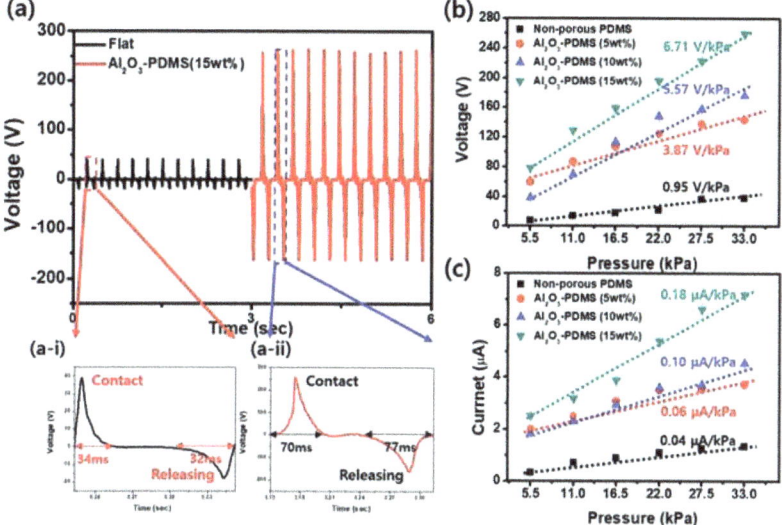

Figure 3. Performance comparison of non-porous PDMS and mesoporous Al$_2$O$_3$-PDMS (5, 10, 15 wt%). (a) The voltage comparison of non-porous PDMS and mesoporous Al$_2$O$_3$-PDMS at 30 N, 4 Hz and 3 × 3 cm^2. The insets show a single peak of (a-i) PDMS and (a-ii) mesoporous Al$_2$O$_3$-PDMS (15 wt%). (b) The voltage and (c) current of mesoporous Al$_2$O$_3$-PDMS (5, 10, 15 wt%) with increasing pressure and the pressure sensitivity.

By analyzing the output voltage and fitting the curve of voltage response to dynamic pressure in the range of 5.5–33 kPa, we determined that the voltage pressure sensitivity of the non-porous PDMS and mesoporous Al$_2$O$_3$-PDMS (5, 10, 15 wt%) reached approximately 0.95, 3.87, 5.57, and 6.71 V/kPa, as shown in Figure 3b. In previous studies, the linearity of pressure sensitivity was divided into several regimes, such as low-pressure and medium-pressure. In this study, however, we found linear pressure sensitivity (sensitivity = 6.37 V/kPa, R^2 = 0.991 in the 5.5–33 kPa pressure region) across the entire pressure regime. This linearity of the sensitivity of mesoporous Al$_2$O$_3$-PDMS is affected by the deformability of the film with the porosity (78%) effect; thus, as porosity increases, the linearity of pressure sensitivity (R^2 value) is high because of the relative capacitance change with thickness change. Like the voltage sensitivity, the current pressure sensitivity reached approximately 0.04, 0.06, 0.10, and 0.18 µA/kPa in Figure 3c, and the linearity was maintained.

Figure 4 shows the schematic and output voltage and current of gapless TENGs, fabricated by attaching an electrode to both sides of mesoporous Al$_2$O$_3$-PDMS film. In gapless TENGs, the output performance was much less than contact-separation mode TENGs because the charge density on the surface had a greater influence than in the pores. Thus, by Relation (1), the total charge density (σ_P) was reduced, and the output voltage and current were also reduced. However, the advantages of gapless devices are simple, allowing for large-scale fabrication, low cost, and good special qualities. In other words, gapless TENGs had very poor output performance, but they neither require complicated fabrication, nor limit the large-scale fabrication of practical systems.

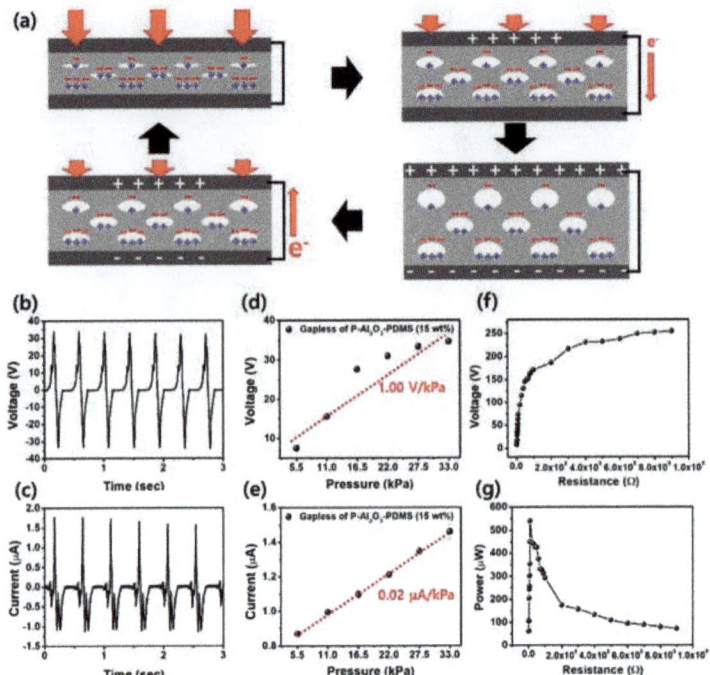

Figure 4. Gapless TENG with a mesoporous Al_2O_3-PDMS (15 wt%). (**a**) The schematic of the mesoporous Al_2O_3-PDMS (15 wt%) gapless system. (**b**) Output voltage and (**c**) current. The pressure sensitivity of the (**d**) voltage and (**e**) current. (**f**) The output voltage and (**g**) power of mesoporous Al_2O_3-PDMS (15 wt%) with resistance of external loads from 10^6–10^9 Ω.

Unlike contact separation mode, charge generation of the mesoporous Al_2O_3-PDMS TENG under pressure is caused by the triboelectric effect and electrostatic induction. First, when mesoporous Al_2O_3-PDMS film is pressurized, friction occurs between PDMS and nano-flake Al_2O_3 MPs in the pores, resulting in negative and positive charges on both surfaces. Furthermore, Al_2O_3 with a nano-flake structure increases charge density due to increasing the surface and friction due to electrostatic induction. With these factors, electrons move from the top electrode to the bottom electrode to keep electrical neutrality. When the pressure is released, the surfaces in contact are separated, and electrons flow from the bottom electrode to the top electrode.

Figure 4b–e shows the output voltage, current, and pressure sensitivity of a mesoporous Al_2O_3-PDMS (15 wt%) gapless system. As mentioned before, gapless systems had output performance inferior to that of contact-separation systems due to no surface charging, and only inner pore charging. At 33.3 kPa, 4 Hz, and 30 × 30 × 3 mm^3, we obtained a maximum voltage of 34.00 V and a current of 1.77 μA. Similar to the contact-separation mode, we measured a pressure sensitivity of 1.00 V/kPa and 0.02 μA/kPa.

The output voltage and power of the mesoporous Al_2O_3-PDMS (15 wt%) were also measured with external loads varying from 1 MΩ–100 MΩ, as shown in Figure 4f,g. It is clearly seen that the output voltage increased with increasing resistance. Consequently, the instantaneous power of the external resistance was 541.11 μW (60.12 μW/cm^2) at a resistance of 10 MΩ.

We also demonstrated the self-powered safety cushion sensor for monitoring the sitting position of the human. Conductive tape arrays (6 × 5), total 30 cells, with each cell size of 3 cm × 3 cm on PET films (33 cm × 27 cm), were fabricated on mesoporous Al_2O_3-PDMS (15 wt%) films, as shown in Figure 5a. To demonstrate the sitting position as represented by color, green was assigned from 1.3–2.5 V, yellow from 2.5–3.8 V, orange from 3.8–5 V, and red over 5 V in real time. We connected the self-powered safety cushion sensor to the multichannel and monitor, as shown in Figure 5b. To confirm the resolution, we sat at the front, in the middle, and towards the back of the safety cushion sensor, as shown in Figure 5c–e. When sitting at the front, only parts of the hips made contact with the safety cushion sensor and only the front of the safety cushion sensor had output voltage. The closer the hips were to the back of the chair, the greater the proportion of safety cushion sensor output voltage, as shown in Figure 5f–h. We are convinced that the self-powered safety cushion sensor could be applied to wheel chairs to control braking if patients sit too far forward.

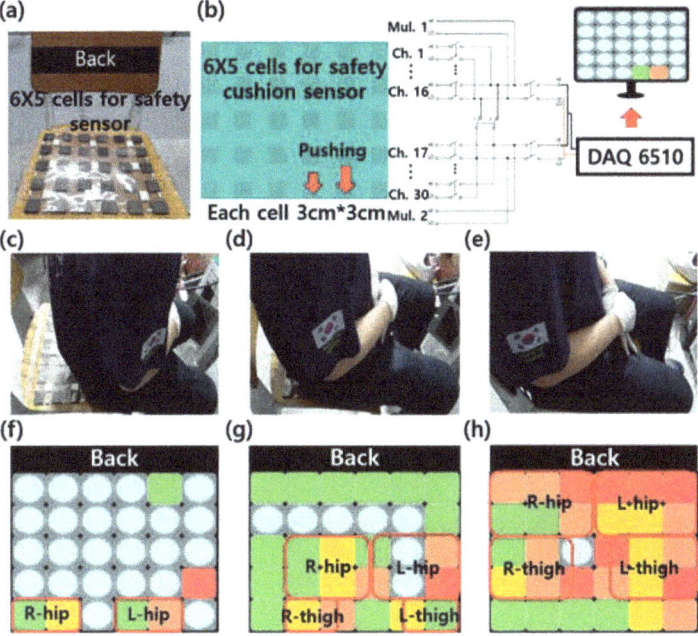

Figure 5. Self-powered safety cushion sensor. (**a**) A photograph of the 6 × 5 cells for the safety sensor, total size of 33 cm × 27 cm; each cell has a size of 3 cm × 3 cm. (**b**) The schematic and circuits of the safety cushion sensor. (**c–e**) Detection of human sitting, front, middle, and back of the sensor array. (**f–h**) The two-dimensional contour plot via monitoring in real time.

4. Conclusions

In summary, we have demonstrated a mesoporous highly-deformable composite polymer with metal oxide by a scalable one-step synthesis unlike the two-step fabrications of previous studies. The oxidation of Al MPs in a polymer film formed a nano-flake morphology on the surface and enhanced the output performance by increasing contact area. As a result, it increased the output performance 6.67-times more than the non-porous PDMS, and the porosity of mesoporous Al_2O_3-PDMS (15 wt%) film reached 71.35%. Thus, mesoporous Al_2O_3-PDMS (15 wt%) film was found to be more deformable and light weight than the non-porous PDMS film and mesoporous films without metal MPs. As a result, we obtained a relative capacitance change of 3.03, about 1.8-times higher compared with mesoporous films without metal MPs at 33.3 kPa. The output performance of mesoporous

Al$_2$O$_3$-PDMS (15 wt%) films showed an instantaneous voltage of 259.68 V under contact-separation mode, an over six-fold enhancement compared with non-porous PDMS. Further, it had a linear pressure sensitivity of 6.71 V/kPa due to the high porosity. We are convinced that the enhancement was affected by the increased charge density created by contact not only between the top electrode and surface of mesoporous Al$_2$O$_3$-PDMS, but also between PDMS and the nano-flake structure on metal MPs' inside pores and high deformation, which changed thickness and increased the electric field. We demonstrated a self-powered safety cushion sensor to detect a human sitting position that can be applied to wheel chair safety. This approach provides a promising large-scale power supply for realizing self-powered safety systems such as matrices, blankets, and road sensors.

Author Contributions: H.J.H. and Y.L. designed the experimental concept and wrote the main manuscript. J.P. discussed the experimental results and commented on the theoretical mechanisms. All authors reviewed the manuscript. The project was guided by D.C. and D.K.

Funding: This study was conducted with the support of the Korea Institute of Industrial Technology (KITECH) as the "Research Source Technique Project" (EO-18-0014) and a National Research Foundation of Korea (NRF) grant funded by the Korean government (MSIP) (No. 2014M3A7B4052202, No. 2017R1A2B2008419, and No. 2016R1A4A1012950).

Conflicts of Interest: The authors declare no conflicts of interest.

References

1. Fan, F.-R.; Tian, Z.-Q.; Wang, Z.L. Flexible triboelectric generator. *Nano Energy* **2012**, *1*, 328–334. [CrossRef]
2. Zhu, G.; Lin, Z.-H.; Jing, Q.; Bai, P.; Pan, C.; Yang, Y.; Zhou, Y.; Wang, Z.L. Toward Large-Scale Energy Harvesting by a Nanoparticle-Enhanced Triboelectric Nanogenerator. *Nano Lett.* **2013**, *13*, 847–853. [CrossRef] [PubMed]
3. Tang, W.; Jiang, T.; Fan, F.R.; Yu, A.F.; Zhang, C.; Cao, X.; Wang, Z.L. Liquid-Metal electrode for High-Performance Triboelectric nanogenerator at an Instantaneous Energy Conversion Efficiency of 70.6%. *Adv. Funct. Mater.* **2015**, *25*, 3718–3725. [CrossRef]
4. Wang, Z.L.; Chen, J.; Lin, L. Progress in triboelectric nanogenerators as a new energy technology and self-powered sensors. *Energy Environ. Sci.* **2015**, *8*, 2250–2282. [CrossRef]
5. Bhatia, D.; Kim, W.; Lee, S.; Kim, S.W.; Choi, D. Tandem triboelectric nanogenerators for optimally scavenging mechanical energy with broadband vibration frequencies. *Nano Energy* **2017**, *33*, 515–521. [CrossRef]
6. Fan, F.-R.; Lin, L.; Zhu, G.; Wu, W.; Zhang, R.; Wang, Z.L. Transparent Triboelectric Nanogenerators and Self-powered Pressure Sensors Based on Micropatterned Plastic Films. *Nano Lett.* **2012**, *12*, 3109–3114. [CrossRef] [PubMed]
7. Lee, K.Y.; Chun, J.; Lee, J.-H.; Kim, K.N.; Kang, N.-R.; Kim, J.-Y.; Kim, M.H.; Shin, K.-S.; Gupta, M.K.; Baik, J.M.; et al. Hydrophobic Sponge Structure-Based Triboelectric Nanogenerator. *Adv. Mater.* **2014**, *26*, 5037–5042. [CrossRef] [PubMed]
8. Xia, X.; Chen, J.; Liu, G.; Javed, M.S.; Wang, X.; Hu, C. Aligning graphene sheets in PDMS for improving output performance of triboelectric nanogenerator. *Carbon* **2017**, *111*, 569–576. [CrossRef]
9. Wang, S.; Xie, Y.; Niu, S.; Lin, L.; Liu, C.; Zhou, Y.S.; Wang, Z.L. Maximum Surface Charge Density for Triboelectric Nanogenerators Achieved by Ionized-Air Injection: Methodology and Theoretical Understanding. *Adv. Mater.* **2014**, *26*, 6720–6728. [CrossRef] [PubMed]
10. Chun, J.; Ye, B.U.; Lee, J.W.; Choi, D.; Kang, J.-Y.; Kim, S.-W.; Wang, Z.L.; Baik, J.M. Boosted output performance of triboelectric nanogenerator via electric double layer effect. *Nat. Commun.* **2016**, *7*, 12985. [CrossRef] [PubMed]
11. Chen, J.; Guo, H.; He, Z.; Liu, G.; Xi, Y.; Shi, H.; Hu, C. Enhancing Performance of Triboelectric Nanogenerator by Filling High Dielectric Nanoparticles into Sponge PDMS film. *ACS Appl. Mater. Interface* **2016**, *8*, 736–744. [CrossRef] [PubMed]
12. Xu, L.; Bu, T.Z.; Yang, X.D.; Zhang, C.; Wang, Z.L. Ultrahigh charge density realized by charge pumping at ambient conditions for triboelectric nanogenerators. *Nano Energy* **2018**, *49*, 625–633. [CrossRef]
13. Mao, Y.; Zhao, P.; McConohy, G.; Yang, H.; Tong, Y.; Wang, X. Sponge-Like Piezoelectric Polymer Films for Scalable and Integratable Nanogenerators and Self-Powered Electronic Systems. *Adv. Energy Mater.* **2014**, *4*, 1301624. [CrossRef]

14. He, X.; Mu, X.; Wen, Q.; Wen, Z.; Yang, J.; Hu, C.; Shi, H. Flexible and transparent triboelectric nanogenerator based on high performance well-ordered porous PDMS dielectric film. *Nano Res.* **2016**, *9*, 3714–3724. [CrossRef]
15. Chun, J.; Kim, J.W.; Jung, W.-S.; Kang, C.-Y.; Kim, S.-W.; Wang, Z.L.; Baik, J.M. Mesoporous pores impregnated with Au nanoparticles as effective dielectrics for enhancing triboelectric nanogenerator performance in harsh environments. *Energy Environ. Sci.* **2015**, *8*, 3006–3012. [CrossRef]
16. Kim, Y.J.; Lee, J.; Park, S.; Park, C.; Park, C.; Choi, H.-J. Effect of the relative permittivity of oxides on the performance of triboelectric nanogenerators. *RSC Adv.* **2017**, *7*, 49368–49373. [CrossRef]

© 2018 by the authors. Licensee MDPI, Basel, Switzerland. This article is an open access article distributed under the terms and conditions of the Creative Commons Attribution (CC BY) license (http://creativecommons.org/licenses/by/4.0/).

Article

Wireless-Powered Chemical Sensor by 2.4 GHz Wi-Fi Energy-Harvesting Metamaterial

Wonwoo Lee [1], Yonghee Jung [1], Hyunseung Jung [2], Chulhun Seo [2], Hosung Choo [3] and Hojin Lee [1,2,*]

[1] Department of ICMC Convergence Technology, Soongsil University, Seoul 06978, Korea; melanie_lee@ssu.ac.kr (W.L.); wjddydg@ssu.ac.kr (Y.J.)
[2] School of Electronic Engineering, Soongsil University, Seoul 06978, Korea; jhs0070@ssu.ac.kr (H.J.); chulhun@ssu.ac.kr (C.S.)
[3] School of Electronic and Electrical Engineering, Hongik University, Seoul 04066, Korea; hschoo@hongik.ac.kr
* Correspondence: hojinl@ssu.ac.kr; Tel.: +82-820-0634

Received: 27 November 2018; Accepted: 24 December 2018; Published: 25 December 2018

Abstract: Metamaterial Sensors show significant potential for applications ranging from hazardous chemical detection to biochemical analysis with high-quality sensing properties. However, they require additional measurement systems to analyze the resonance spectrum in real time, making it difficult to use them as a compact and portable sensor system. Herein, we present a novel wireless-powered chemical sensing system by using energy-harvesting metamaterials at microwave frequencies. In contrast to previous studies, the proposed metamaterial sensor utilizes its harvested energy as an intuitive sensing indicator without complicated measurement systems. As the spectral energy-harvesting rate of the proposed metamaterial sensor can be varied by changing the chemical components and their mixtures, we can directly distinguish the chemical species by analyzing the resulting output power levels. Moreover, by using a 2.4 GHz Wi-Fi source, we experimentally realize a prototype chemical sensor system that wirelessly harvests the energy varying from 0 mW up to 7 mW depending on the chemical concentration of the water-based binary mixtures.

Keywords: energy-harvesting metamaterial; wireless chemical sensor; metamaterial sensor

1. Introduction

Metamaterials are artificially engineered structures that show exotic electromagnetic properties including strong field enhancement and localization of the incident waves [1,2], which consequently offer high-quality resonances and sensing abilities [3–5]. With these advantages, metamaterial-based sensor systems have attracted considerable attention for highly sensitive and nondestructive detection applications such as temperature [6], food quality [7–9], humidity [10], chemical [11–16], and biological sensor systems [17–21]. In particular, metamaterial-based chemical sensor systems are increasingly necessary for use as an environment sensor where immediate and accurate detection of various chemical substances is required for human safety. However, in real-time applications, most of the conventional metamaterial-based sensor systems are limited by the requirement of external equipments such as a network analyzer [14–16], spectroscopy systems [9,16–19], and liquid pump systems [12–15].

To overcome these problems, energy-harvesting metamaterials can be a good candidate for a new sensing platform because their spectral resonance properties can be extracted from the harvested energies. Energy-harvesting metamaterials have been recently studied for the conversion of external electromagnetic energy into a direct current (DC) signal to supply electric power to diverse wireless power systems [22–26] and wireless sensor networks [27–29]. As the resonance properties of the metamaterials are sensitive to their environment [15,17,18], we can expect that the maximum harvesting rate can be sensitively changed by various chemical substances dropped on the energy harvesting metamaterials. Therefore, by plotting the variation of the energy harvested from the specific incident waves, the change in the surrounding environment owing to the change in the concentration or species of chemical substances can be analyzed.

In this paper, we proposed a novel self-powered energy-harvesting metamaterial sensor system operating at microwave frequencies. A commercially available 2.4 GHz Wi-Fi router with an antenna was used as a power source to supply electromagnetic waves wirelessly to our metamaterial sensor. In our sensor design, a micro-channel was constructed at the gap of the metamaterial pattern to confine the chemicals in the area of the enhanced electric field to effectively detect minute amounts of chemicals [15]. The chemical components of the various solutions could be identified by the metamaterial-based sensor through the analysis of the energy-harvesting rate of our sensor system, which was varied by changing the concentration and species of chemicals. In contrast to the previously reported metamaterial sensor systems, by converting the harvested free microwave energy into DC current, the proposed metamaterial sensor could exhibit satisfactory sensing results with simple LED indicators without requiring complicated measurement setup and additional input energy sources.

2. Materials and Methods

2.1. Design of Energy-Harvesting Metamaterial Sensor

Figure 1a illustrates the schematic view of the proposed energy-harvesting metamaterial sensor, which consists of a single split-ring resonator (SRR), chemical channel, and rectifier circuit for converting the incident microwave into DC output voltage. As the SRR structure can be considered an LC circuit consisting of an inductive metallic loop and capacitive gap, an electric field is strongly induced only at the SRR gap area at the resonance frequency, as shown in Figure 1b. Hence, if the chemical components are captured within the channel placed in this gap, the resonance frequency of the metamaterials can be effectively changed, because the electrical permittivity of the chemical components affects the effective capacitance around the gap of the SRR resonator. Simultaneously, to operate our metamaterial structure as an energy-harvesting device, a Greinacher voltage doubler circuit [22] was adopted and connected to two output nodes of SRR to rectify the induced current in the SRR structure by the incident microwave.

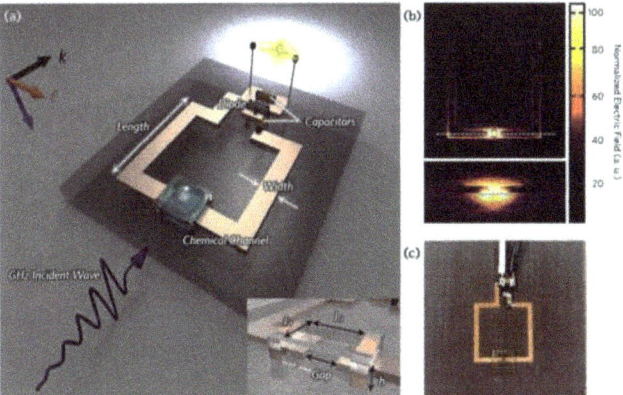

Figure 1. Schematics of the proposed system and design. (**a**) Schematic of the proposed energy-harvesting metamaterial sensor system. (**b**) Top and cross-sectional views of the simulated electric field distribution of the proposed metamaterial device at the resonance frequency. (**c**) Photograph of the fabricated metamaterial sensor system.

A geometrical structure of the metamaterial sensor was designed to operate in the range approximately 2 GHz to 3 GHz with the resonance frequency of 2.4 GHz using ANSYS high-frequency structure simulator (HFSS). An SRR made of copper pattern was fabricated on a Taconic TLC substrate with the permittivity of $\varepsilon = 3.2$ using photolithography. Based on the simulation results, the geometry of the SRR was optimized to have the length (l) of 13 mm, width (w) of 1 mm, and gap (g) of 1 mm. The rectifier circuit was connected to the electrodes of the SRR loop. The chemical channel was made of SU-8 photoresist owing to its good chemical resistance [30,31].

2.2. Sample Fabrication

The proposed energy-harvesting metamaterial chemical sensor was fabricated by the following process. First, a 35-µm-thick copper on 0.78-mm-thick Taconic TLC substrate was patterned using photolithography process for single split-ring resonator (SRR) pattern. After patterning SRR, to fabricate chemical channel, SU-8 100 photoresist (Microchem Corporation, Westborough, MA, USA) was spin-coated on copper-patterned substrate and soft-baked at 70 °C for 5 min followed by 100 °C for 100 min. Then, the photoresist-coated substrate was exposed to UV aligner using photomask. After exposure, the substrate was annealed at 70 °C for 1 min followed by 100 °C for 35 min. Then, by dissolving the substrate into the SU-8 developer (Microchem Corporation), the chemical channel was successfully formed. Finally, for the rectifier circuit, the diode (HSMS-2862-BLGK, San Jose, CA, USA) and chip capacitors were connected by soldering. As shown in Figure 1c, the chemical channel was formed at the SRR gap to detect the chemical components by using a simple drop casting method. The chemical channel was designed with the dimensions $l_1 = 300$ µm, $l_2 = 500$ µm, and $h = 600$ µm. The fabricated metamaterial sensor device is shown in Figure 1c with a transparent chemical channel covering the SRR gap area.

2.3. Water-based Bianry Mixture Chemical Components Preparation

Water-based binary mixture chemical components were prepared by mixing deionized water (DI water) with solution-state chemicals such as toluene, chloroform, ethanol, acetone, and methanol with a purity of 99.5% or higher. To control the exact concentration and the precise amount of chemicals, a micropipette was used to extract DI water and chemicals. For the calculation of concentration, volume percent (v/v % chemical) was used. For example, in order to get 100 mL acetone chemical components with 20% concentration, 20 mL of acetone solution was dissolved in 80 mL DI water solution.

3. Results

3.1. Spectral Resonance Frequency Shift of Energy-Harvesting Metamaterial Sensor

Figure 2 shows the schematics of sample preparation by drop casting and structures of sensing chemical materials. Chemical components are dropped into the micro-channel using micropipette to accurately control the amounts of chemicals. Since this micro-channel around the SRR gap minimizes geometrical variation factors of the liquid chemical components, we can evaluate that the resonance changes of our metamaterial sensors solely originated from the dielectric constant of the chemical components.

To determine how effectively the chemical substance at the SRR gap affected the spectral resonance of our metamaterial sensor, we simulated the transmission spectra of the metamaterial sensor for various chemical substances with different dielectric constants, and confirmed that the resonance frequency of transmission of the metamaterial sensor was shifted by the use of different species of chemical components having different dielectric constants as shown in Figure 3. To verify the simulation results, we filled the chemical channel of the metamaterial sensor with various chemical substances such as toluene ($\varepsilon_t = 2.38$) [32], chloroform ($\varepsilon_c = 4.81$) [33], ethanol ($\varepsilon_e = 9$) [34], acetone ($\varepsilon_a = 21$) [35], methanol ($\varepsilon_m = 33.1$) [34], and water ($\varepsilon_w = 79.5$) [34], and measured the output voltage (V_{out}) of the metamaterial sensor in the frequency domain. To measure the spectral resonance properties of the energy-harvesting metamaterial sensor, a horn antenna was connected to the signal generator to generate an incident electromagnetic wave and the metamaterial sensor was placed at a distance of 10 cm from the horn antenna. Then, spectral output voltages and resonance properties [36–38] of the proposed metamaterial sensor were measured for different frequencies of the incident wave ranging from 1.8 GHz to 2.7 GHz and all of the measurement system was surrounded by anechoic materials. As shown in Figure 3, the measured output voltage showed distinct resonance frequency shifts for the various chemical components as confirmed by the simulation results. When toluene, chloroform, ethanol, acetone, methanol, and water were captured at the chemical channel, the resonance peak of the metamaterial sensor was shifted to 2.32, 2.28, 2.20, 2.08, 1.97, and 1.78 GHz, respectively.

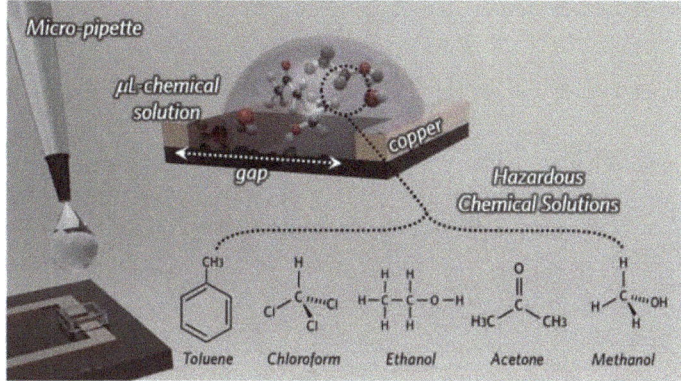

Figure 2. Schematic of sample preparation by drop casting and structures of sensing chemical materials.

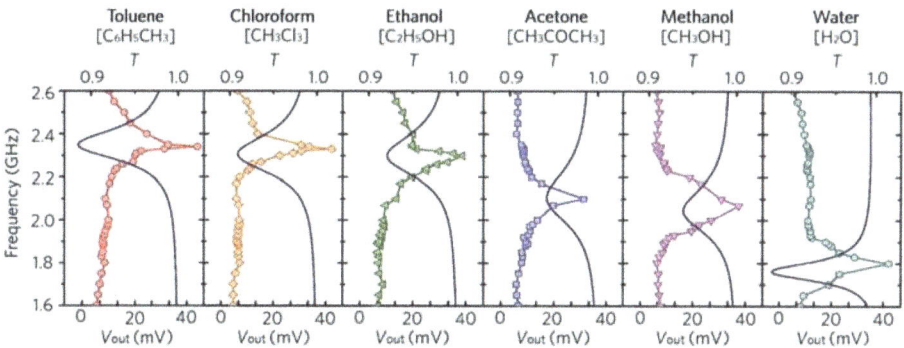

Figure 3. Simulated transmission spectra (line) and measured output voltage (V_{out}) spectral (symbol) of the proposed metamaterial sensor system for various chemical substances.

To investigate the chemical sensing properties of our proposed system, we experimentally evaluated various chemical mixtures, as shown in Figure 4. As the effective dielectric constant for chemical mixtures can be calculated from the Braggman Equation [39], we plotted the calculated effective dielectric constant for binary chemical mixtures, as shown in Figure 4a. Owing to the discrepancy of the dielectric constant between water and chemical mixtures, the effective dielectric constants can be varied by changing the chemical mixture ratios and concentrations. Based on the calculated effective dielectric constants, we performed a simulation to observe the change in the resonance frequency and confirmed that the resonance frequency shifted sensitively as the concentration changed. An experiment was performed to demonstrate this, and results similar to the simulation results were obtained. Therefore, as shown in Figure 4b,c, the simulated and measured resonance frequencies of the metamaterial sensor were gradually blue-shifted in accordance with the increase in the chemical concentration in the water-based binary mixtures.

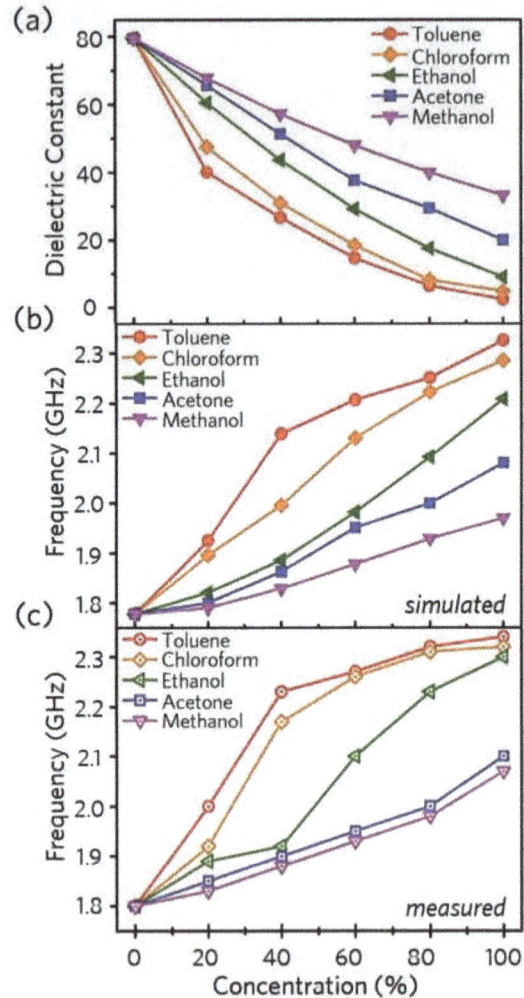

Figure 4. (a) Calculated effective dielectric constants for the chemical mixtures. (b) Simulated and (c) measured resonance frequencies for the chemical mixtures.

3.2. Wi-Fi Energy-Harvesting Metamaterial Sensor System

For practical uses, we adopted an easily accessible and widespread 2.4 GHz Wi-Fi access-point (AP) as an electromagnetic single wave source for our wireless sensing system. The Wi-Fi AP was placed at a distance of 10 cm from the metamaterial sensor and a light-emitting diode (LED) was connected to the metamaterial sensor as an indicator to evaluate the energy-harvesting rate, as shown in Figure 5a. To compare the energy-harvesting rate depending on the resonance properties, we fabricated several devices with different resonance frequencies ranging from 2 GHz to 3 GHz and measured the DC output voltage of the sensors, which indicates the energy harvested from the Wi-Fi sources. As shown in Figure 5b, the highest output voltage of 3.38 V was obtained for the metamaterial device with the resonance frequency of 2.4 GHz, because the resonance frequency of the sensor was matched with the Wi-Fi source frequency. The energy-harvesting rate decreased to 1.5 V when the resonance frequency of the metamaterial sensor shifted to lower or higher frequencies from 2.4 GHz. Thus, we could confirm that the difference in the energy-harvesting rate was due to the discrepancy between the resonance frequencies of the Wi-Fi source and the metamaterial sensor. Hence, by using the resonance frequency shift of the metamaterial sensor for the different chemicals, we can identify the chemical substances and mixture ratios by plotting the variation of the energy-harvesting rate of the metamaterial sensor. Figure 6a shows the measured output voltages of various chemical mixtures with concentrations varying from 0 % to 100 % in increments of 20 %. The maximum output DC voltage of 2.98 V could be obtained for 100 % toluene when the resonance frequency of the device was 2.34 GHz as shown in Figure 6a. When the ratio of water in the chemical mixtures was increased, the output voltage decreased and the minimum output voltage was measured at around 560 mV for 100 % water. To evaluate the chemical sensitivity of the proposed sensor more intuitively, an LED was adopted as an indicator whose luminescence was proportional to the harvested energy determined by the chemical components in the binary mixtures. As the change in the dielectric constant of the chemical compounds determines the energy-harvesting rate, the resulting output power varied from 0 mW to 7 mW depending on the chemical substances and their mixture ratios, and the brightness of the LED changed correspondingly, as shown in Figure 6b,c. Figure 6c shows the experimental results of the LED-embedded metamaterial harvesting sensor system for different chemical concentrations in the water-based binary mixtures. Hence, as the energy-harvesting rate of the sensor for toluene, chloroform, and ethanol binary mixtures decreased according to the increase in the water concentration (Figure 6a), the brightness of the LED also tended to decrease for all the chemical mixtures, and was eventually completely turned off for 100 % water concentration as the output voltage of the sensor was smaller than the turn-on voltage of the LED.

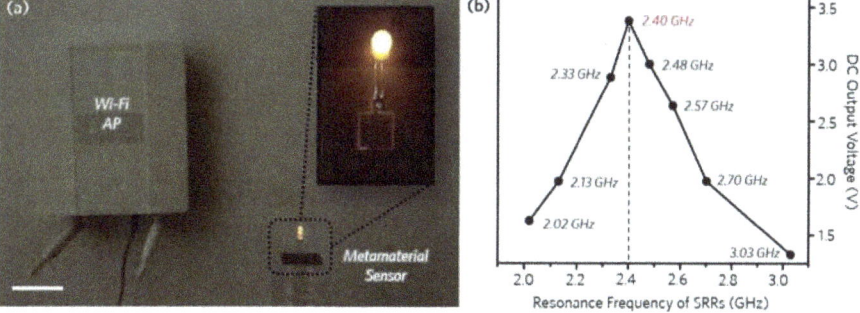

Figure 5. Wi-Fi energy-harvesting sensor system. (**a**) Photograph of the experimental setup for the Wi-Fi energy-harvesting sensor sys-tem. The inset shows the energy-harvesting sensor connected to an LED. (scale bar: 5 cm) (**b**) Measured DC output voltage of the Wi-Fi harvesting metamaterials with different resonance frequencies.

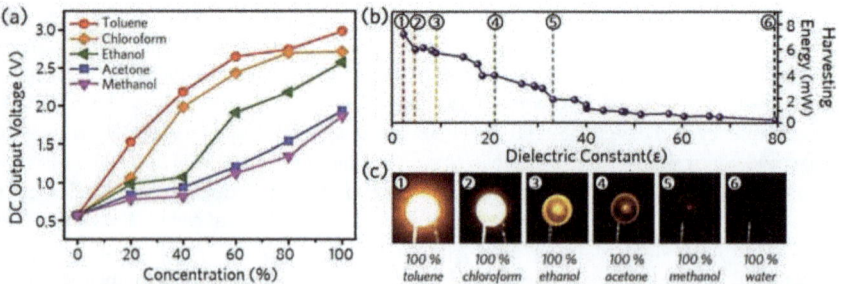

Figure 6. Experimental results for various chemical mixtures. (**a**) Measured DC output voltage of the Wi-Fi harvesting sensor for various concentrations of the chemical mixtures. (**b**) Measured harvesting energy and (**c**) corresponding brightness of the sensor for different dielectric constants of the chemical mixtures.

4. Conclusions

In summary, we presented a novel wireless chemical sensor system based on energy-harvesting metamaterials at microwave frequencies. The change in the effective dielectric constants with different chemical compounds and their mixtures led to the resonance shift of our metamaterial sensor and thus resulted in the spectral change of its energy-harvesting properties. From the simulation and experimental results, we confirmed that the proposed sensor system successfully realized wireless chemical sensing by utilizing a commercial 2.4 GHz Wi-Fi AP as a single wave source without any external power source. Moreover, as the maximum energy-harvesting rate occurred at the resonance frequency of our metamaterial device, the proposed sensor system could provide simple LED-based chemical analysis by measuring the chemical-dependent energy-harvesting rates without additional complicated measurement systems, thus, offering accessibility and simplicity for sensor applications. We believe that our results can pave the way for miniaturized wireless sensor systems, including biochemical and dielectric environment detectors.

Author Contributions: W.L. and Y.J. contributed equally. W.L. and Y.J. designed and fabricated the metamaterial sensors. H.J. simulated and analyzed the metamaterial structure and circuit design. H.C. prepared microwave measurement set-up. H.L. guided the research. W.L., Y.J., H.J., C.S., and H.L. prepared the manuscript and contributed to the data analysis and discussions.

Funding: This work was supported by the National Research Foundation of Korea (NRF) grant funded by the Korean government (MSIP)(NRF-2017R1A5A1015596).

Conflicts of Interest: The authors declare no conflict of interest.

References

1. Seo, M.A.; Park, H.R.; Koo, S.M.; Park, D.J.; Kang, J.H.; Suwal, O.K.; Choi, S.S.; Planken, P.C.M.; Park, G.S.; Park, N.K.; et al. Terahertz field enhancement by a metallic nano slit operating beyond the skin-depth limit. *Nat. Photonics* **2009**, *3*, 152–156. [CrossRef]
2. Chen, W.T.; Chen, C.J.; Wu, P.C.; Sun, S.; Zhou, L.; Guo, G.-Y.; Hsiao, C.T.; Yang, K.-Y.; Zheludev, N.I.; Tsai, D.P. Optical magnetic response in three-dimensional metamaterial of upright plasmonic meta-molecules. *Opt. Express* **2011**, *19*, 12837–12842. [CrossRef] [PubMed]
3. Cubukcu, E.; Zhang, S.; Park, Y.-S.; Bartal, G.; Zhang, X. Split ring resonator sensors for infrared detection of single molecular monolayers. *Appl. Phys. Lett.* **2009**, *95*, 043113. [CrossRef]
4. Chang, Y.-T.; Lai, Y.-C.; Li, C.-T.; Chen, C.-K.; Yen, T.-J. A multi-functional plasmonic biosensor. *Opt. Express* **2010**, *18*, 9561–9569. [CrossRef] [PubMed]
5. Chen, T.; Li, S.; Sun, H. Metamaterials application in sensing. *Sensors* **2012**, *12*, 2742–2765. [CrossRef] [PubMed]

6. Kim, H.; Delfin, D.; Shuvo, M.A.I.; Chavez, L.A.; Carcia, C.R.; Barton, J.H.; Gaytan, S.M.; Cadena, M.A.; Rumpf, R.C.; Wicker, R.B.; et al. Concept and Model of a Metamaterial-Based Passive Wireless Temperature Sensor for Harsh Environment Applications. *IEEE Sens. J.* **2015**, *15*, 1445–1452. [CrossRef]
7. Tao, H.; Brenckle, M.A.; Yang, M.; Zhang, J.; Liu, M.; Siebert, S.M.; Averitt, R.D.; Mannor, M.S.; McAlpine, M.C.; Rogers, J.A.; et al. Silk-Based Conformal, Adhesive, Edible Food Sensors. *Adv. Mater.* **2012**, *24*, 1067–1072. [CrossRef]
8. Potyrailo, R.A.; Nagraj, N.; Tang, Z.; Mondello, F.J.; Surman, C.; Morris, W. Battery-free Radio Frequency Identification (RFID) Sensors for Food Quality and Safety. *J. Agric. Food Chem.* **2012**, *60*, 8535–8543. [CrossRef]
9. Tamer, A.; Alkurt, F.; Altintas, O.; Karaaslan, M.; Unal, E.; Akgol, O.; Karadag, F.; Sabah, C. Transmission Line Integrated Metamaterial Based Liquid Sensor. *J. Electrochem. Soc.* **2018**, *7*, B251–B257. [CrossRef]
10. Amin, E.M.; Bhuiyan, M.S.; Karmakar, N.C.; Winther-Jensen, B. Development of a Low Cost Printable Chipless RFID Humidity Sensor. *IEEE Sens. J.* **2014**, *14*, 140–149. [CrossRef]
11. Wu, P.C.; Sun, G.; Chen, W.T.; Yang, K.-Y.; Huang, Y.-W.; Chen, Y.-H.; Huang, H.L.; Hsu, W.-L.; Chiang, H.P.; Tsai, D.P. Vertical split-ring resonator based nanoplasmonic sensor. *Appl. Phys. Lett.* **2014**, *105*, 033105. [CrossRef]
12. Withayachumnankul, W.; Jaruwongrungsee, K.; Tuantranont, A.; Fumeaux, C.; Abbott, D. Metamaterial-based microfluidic sensor for dielectric characterization. *Sens. Actuators A Phys.* **2013**, *189*, 233–237. [CrossRef]
13. Ebrahimi, A.; Withayachumnankul, W.; Al-Sarawi, S.; Abbott, D. High-sensitivity metamaterial-inspired sensor for microfluidic dielectric characterization. *IEEE Sens. J.* **2014**, *14*, 1345–1351. [CrossRef]
14. Yoo, M.; Kim, H.K.; Lim, S. Electromagnetic-based ethanol chemical sensor using metamaterial absorber. *Sens. Actuators B Chem.* **2016**, *222*, 173–180. [CrossRef]
15. Shih, K.; Pitchappa, P.; Manjappa, M.; Ho, C.P.; Singh, R.; Lee, C. Microfluidic metamaterial sensor: Selective trapping and remote sensing of microparticles. *J. Appl. Phys.* **2017**, *121*, 023102. [CrossRef]
16. Bakir, M.; Karaaslan, M.; Dincer, F.; Delihacioglu, K.; Sabah, C. Perfect metamaterial absorber-based energy harvesting and sensor applications in the industrial, scientific, and medical band. *Opt. Eng.* **2015**, *54*, 097102. [CrossRef]
17. Anker, J.N.; Hall, W.P.; Lyandres, O.; Shah, N.C.; Zhao, J.; Duyne, R.P.V. Biosensing with plasmonic nanosensors. *Nat. Mater.* **2008**, *7*, 442–453. [CrossRef]
18. Park, S.J.; Hong, J.T.; Choi, S.J.; Kim, H.S.; Park, W.K.; Han, S.T.; Park, J.Y.; Lee, S.; Kim, D.S.; Ahn, Y.H. Detection of microorganisms using terahertz metamaterials. *Sci. Rep.* **2014**, *4*, 4988. [CrossRef]
19. Xie, L.; Gao, W.; Shu, J.; Ying, Y.; Kono, J. Extraordinary sensitivity enhancement by metasurfaces in terahertz detection of antibiotics. *Sci. Rep.* **2015**, *5*, 8671. [CrossRef]
20. Lee, H.-J.; Yook, J.-G. Biosensing using split-ring resonators at microwave regime. *Appl. Phys. Lett.* **2008**, *92*, 254103. [CrossRef]
21. Lee, H.-J.; Lee, J.-H.; Moon, H.-S.; Jang, I.-S.; Choi, J.-S.; Yook, J.-G.; Jung, H.-I. A planar split-ring resonator-based microwave biosensor for label-free detection of biomolecule. *Sens. Actuators B Chem.* **2012**, *169*, 26–31. [CrossRef]
22. Hawkes, A.M.; Katko, A.R.; Cummer, S.A. A microwave metamaterial with integrated power harvesting functionality. *Appl. Phys. Lett.* **2013**, *103*, 163901. [CrossRef]
23. Wang, B.; Teo, K.H.; Nishino, T.; Yerazunis, W.; Barnwell, J.; Zhang, J. Experiments on wireless power transfer with metamaterial. *Appl. Phys. Lett.* **2011**, *98*, 254101. [CrossRef]
24. Huang, D.; Urzhumov, Y.; Smith, D.R.; Teo, K.H.; Zhang, J.Y. Magnetic superlens-enhanced inductive coupling for wireless power transfer. *J. Appl. Phys.* **2012**, *111*, 064902. [CrossRef]
25. Ranaweera, A.; Duong, T.P.; Lee, J.-W. Experimental investigation of compact metamaterial for high efficiency mid-range wireless power transfer applications. *J. Appl. Phys.* **2014**, *116*, 043914. [CrossRef]
26. Lipworth, G.; Ensworth, J.; Seetharam, K.; Huang, D.; Lee, J.S.; Schmalenberg, P.; Nomura, T.; Reynolds, M.S.; Smith, D.R.; Urzhumov, Y. Magnetic Metamaterial Superlens for Increased Range Wireless Power Transfer. *Sci. Rep.* **2014**, *4*, 3642. [CrossRef]
27. Alippi, C.; Galperti, C. An Adaptive System for Optimal Solar Energy Harvesting in Wireless Sensor Network Nodes. *IEEE Trans. Circuits Syst. I Reg. Pap.* **2008**, *55*, 1742–1750. [CrossRef]
28. Nishimoto, H.; Kawahara, Y.; Asami, T. Prototype Implementation of Ambient RF Energy Harvesting Wireless Sensor Networks. In Proceedings of the IEEE Sensors, Kona, HI, USA, 1–4 November 2010.

29. Kong, H.-B.; Wang, P.; Niyato, D.; Cheng, Y. Modeling and Analysis of Wireless Sensor Networks with/without Energy Harvesting Using Ginibre Point Processes. *IEEE Trans. Wirel. Commun.* **2017**, *16*, 3700–3713. [CrossRef]
30. Balslev, S.; Jorgensen, A.M.; Bilenberg, B.; Mogensen, K.B.; Snakenborg, D.; Geschke, O.; Kutter, J.P.; Kristensen, A. Lap-on-a-chip with integrated optical transducers. *Lab Chip* **2006**, *6*, 213–217. [CrossRef] [PubMed]
31. Nordström, M.; Marie, R.; Calleja, M.; Boisen, A. Rendering SU-8 hydrophilic to facilitate use in micro channel fabrication. *J. Micromech. Microeng.* **2004**, *14*, 1614–1617. [CrossRef]
32. Lide, D.R. *Handbook of Chemistry and Physics*, 72nd ed.; CRC Press: Boca Raton, FL, USA, 1992; pp. 1229–1379. ISBN 978-0-849-30472-9.
33. Cataldo, A.; Benedetto, E.D.; Cannazza, G. *Broadband Reflectometry for Enhanced Diagnostics and Monitoring*; Spinger: Berlin/Heidlberg, Germany, 2011; p. 79. ISBN 978-3-642-20233-9.
34. Bao, J.; Swicord, M.L.; Davis, C.C. Microwave dielectric characterization of binary mixtures of water, methanol, and ethanol. *J. Chem. Phys.* **1996**, *104*, 4441–4450. [CrossRef]
35. Onimisi, M.Y.; Ikyumbur, J.T.; Abdu, S.G.; Hemba, E.C. Frequency and Temperature Effect on Dielectric Properties of Acetone and Dimethylformamide. *PSIJ* **2016**, *11*, 1–8. [CrossRef]
36. Harouni, Z.; Cirio, L.; Osman, L.; Gharsallah, A.; Picon, O. A Dual Circularly Polarized 2.45-GHz Rectenna for Wireless Power Transmission. *IEEE Antennas Wirel. Propag. Lett.* **2011**, *10*, 306–309. [CrossRef]
37. Keyrouz, S.; Visser, H. Efficient Direct-Matching Rectenna Design for RF Power Transfer Applications. *J. Phys. Conf. Ser.* **2013**, *476*, 012093. [CrossRef]
38. Heikkinen, J.; Kivikoski, M. Low-Profile Circularly Polarized Rectifying Antenna for Wireless Power Transmission at 5.8 GHz. *IEEE Microw. Compon. Lett.* **2004**, *14*, 162–164. [CrossRef]
39. Zhang, D.; Cherkaev, E.; Lamoureux, M.P. Stieltjes representation of the 3D Bruggeman effective medium and Padé approximation. *Appl. Math. Comput.* **2011**, *217*, 7092–7107. [CrossRef]

© 2018 by the authors. Licensee MDPI, Basel, Switzerland. This article is an open access article distributed under the terms and conditions of the Creative Commons Attribution (CC BY) license (http://creativecommons.org/licenses/by/4.0/).

Article

Cylindrical Free-Standing Mode Triboelectric Generator for Suspension System in Vehicle

Minki Kang [1], Tae Yun Kim [1], Wanchul Seung [1], Jae-Hee Han [2] and Sang-Woo Kim [1,*]

[1] School of Advanced Materials Science and Engineering, Sungkyunkwan University (SKKU), Suwon 16419, Korea; kang0824@skku.edu (M.K.); herotyun@skku.edu (T.Y.K.); s2000@skku.edu (W.S.)
[2] Department of Energy IT, Gachon University, Seongnam 13120, Korea; jhhan388@gachon.ac.kr
* Correspondence: kimsw1@skku.edu; Tel.: +82-31-290-7632

Received: 20 November 2018; Accepted: 18 December 2018; Published: 29 December 2018

Abstract: The triboelectric generator (TEG) is a strong candidate for low-power sensors utilized in the Internet of Things (IoT) technology. Within IoT technologies, advanced driver assistance system (ADAS) technology is included within autonomous driving technology. Development of an energy source for sensors necessary for operation becomes an important issue, since a lot of sensors are embedded in vehicles and require more electrical energy. Although saving energy and enhancing energy efficiency is one of the most important issues, the application approach to harvesting wasted energy without compromising the reliability of existing mechanical systems is still in very early stages. Here, we report of a new type of TEG, a suspension-type free-standing mode TEG (STEG) inspired from a shock absorber in a suspension system. We discovered that the optimum width of electrode output voltage was 131.9 V and current was 0.060 µA/cm^2 in root mean square (RMS) value while the optimized output power was 4.90 µW/cm^2 at 66 MΩ. In addition, output power was found to be proportional to frictional force due to the contact area between two frictional surfaces. It was found that the STEG was made of perfluoroalkoxy film and showed good mechanical durability with no degradation of output performance after sliding 11,000 times. In addition, we successfully demonstrated charging a capacitor of 330 µF in 6 min.

Keywords: triboelectric generator; shock absorber; suspension system; advanced driver assistance technology; IoT technology; frictional force

1. Introduction

Internet of Things (IoT) technology is becoming more influential within our lives recently. This technology, which is based on wireless sensor network systems, has comprehensive applications, such as health monitoring, smart factories, autonomous driving, etc. Among them, autonomous driving is one of the most promising technologies. As an intermediate stage, a number of automotive companies are trying to develop advanced driver assistance system (ADAS) technology to realize autonomous driving vehicles. As ADAS technology is based on sensor network systems, continuous power supply for stable and long-term operation of sensors is very important. Although individual sensors for sensor networks consume relatively small amounts of energy, ranging from µW to mW, large amounts of energy would be required for the sustainable operation of a huge number of sensors assembled in an IoT system [1,2]. However, current automotive energy harvesting technologies based on traditional electro-magnetic induction are not efficient for powering sensors because of their heavy weight when compared to their output performance, and thus the absence of an efficient energy source for the sensor network system, not only leads to a reduction in energy efficiency of the vehicle, but also disturbs the practical application of ADAS technology in a vehicle [3,4].

Triboelectric generators (TEGs) that convert mechanical energy to electrical energy utilize a combination of triboelectrification and electrostatic induction, which is one of the most promising candidates as an energy source of low-power sensors because of its high output performance, light weight, and low cost [5–8]. Based on these attributes, there have been numerous attempts to harvest mechanical energy which is abandoned in vehicle such as energy from wind, vibration, rotation and etc through applying TEGs [9–11]. However, TEGs that rely entirely on mechanical friction may deteriorate the energy efficiency of vehicles by causing more energy consumption, which is required for driving. For example, a rotating free-standing mode TEG applied to the wheel makes a vehicle consume more energy due to the continuous mechanical friction-induced driving resistance. Alternatively, when a TEG harvesting wind energy blows it into a vehicle, it results in high air resistance that disturbs driving, and thus consumes more energy [9–11]. Therefore, it is an important issue to design appropriate structures and harvest inevitably-wasted energy without reducing energy conversion efficiency and the reliability of the vehicles.

Here, we suggest that suspension-type free-standing mode TEG (STEG) can be applicable to reciprocal movement of a shock absorber in the suspension system of vehicles. The frictional force of STEG can perfectively replace the damping force of a conventional shock absorber and does not cause additional energy consumption. STEG has a novel, cylindrical, free-standing mode TEG structure that consists of an aluminum inner cylinder and outer cylinder. The triboelectric charges generated by the friction between a perfluoroalkoxy (PFA) film and grating-structured outer cylinder induce electrostatic induction to the alternating two electrodes on the inner cylinder and generate displacement current. We validated output performance of the STEG, which has a 1:8 scaled size of a real shock absorber in the vehicles, as a function of width of the electrode and speed of the reciprocal movement. In previous reports, many factors that affect the generating performance of TEGs have been investigated, such as charge density, permittivity, thickness of film, and surface area. However, the frictional force between two different materials in sliding mode STEGs have not yet been considered because of the complexity of phenomenon and difficulties in control. Frictional force is proportional to normal force, and the effective contact area increases as the normal force becomes stronger. We obtained the highest output performance of STEG at 4.90 $\mu W/cm^2$ by controlling the frictional force with different radii of the inner cylinder.

2. Materials and Methods

2.1. The Fabrication Process of the STEG

The STEG is comprised of an inner cylinder and an outer cylinder made of aluminum with a 1:8 scale sized shock absorber in vehicles. The diameter of the inner cylinder was 23 mm and was surrounded by copper electrodes deposited on a flexible printed circuit board (PCB) (substrate is polyimide, 12.5 µm thick). The copper electrodes (12.5 µm thick) had an interdigitated structure with a 1-mm gap between adjacent electrodes in order to prevent short circuits, and were deposited on the same area of 70 × 130 mm^2 to investigate the optimum design. Polyethylene (PE) foam tape (1 mm thick) was inserted between the inner cylinder and the surrounding copper electrode-deposited PCB to increase the outer diameter of the inner cylinder and control the frictional force between the inner cylinder and the outer cylinder. Lastly, PFA film, which is a frictional layer with a thickness of 25 µm, was attached to the copper electrode using commercial double-sided tape. The inner diameter of the outer cylinder is 26 mm. The inner surface of outer cylinder which is facing the PFA film on the inner cylinder has a grating structure with the same width as the copper electrode. Both the inner and outer cylinders have lengths of 220 mm.

2.2. Characterization of STEG

Output performance of STEG was measured while the inner surface of the outer cylinder slides on the PFA film surface by applying a cyclic vertical force using pushing tester (ZPS-100, JUNIL TECH Co., Ltd., Seoul, Korea). The inner cylinder was fixed on the bottom stage and the outer cylinder was attached to the moving part, which periodically moved with a constant speed of 62.5, 140, and 200 mm/s, respectively, and the period was 0.4 s for all cases. During movement, the frictional force controlled by PE foam tape was measured using a force sensor installed in the pushing tester. The output voltage between two interdigitated electrodes was measured using an oscilloscope (DPO 3052, Tektronix, Beaverton, OR, USA) with an input impedance of 40 MΩ, and the output current was measured using a current amplifier (DLPCA-200, FEMTO, Berlin, Germany) connected to the oscilloscope under short circuit conditions. After 11,000 periodic movements with a frictional force of 0.6 kgf, the surface morphology of PFA film was examined using scanning electron microscopy (JSM-6701F, FE-SEM, Jeol Ltd., Mitaka, Tokyo, Japan) to check mechanical durability.

3. Results

3.1. Geometrical Design of the STEG and Electrical Performance

The STEG was designed based on the structure of a free-standing mode TEG, in which the copper electrodes on the inner cylinder and outer cylinder corresponded to alternative electrodes and moving objects, respectively. For the investigation and fabrication of a STEG that is compatible with the current suspension systems in vehicles, the materials were selected to satisfy, not only the triboelectric series, which indicates how much the material has tendency to have a positive or negative surface charge, but also to maintain the reliability of the suspension system.

The schematic structure and triboelectric energy regenerative suspension application is featured in Figure 1a. Each cylinder was made of aluminum, and flexible PCB comprised of polyimide (PI) substrate and alternating copper electrodes was covered by a perfluoroalkoxy film (PFA) on the inner cylinder. Optical images, material, and structure of STEG are shown in Figure S1. The outer cylinder had a grating structure on the inner surface and a slide along the PFA film, generating triboelectric charges. The size of the cylinder was 1:8 scaled size compared to commercial shock absorbers in the suspension systems of vehicles.

The working mechanism of STEG is based on that of a free-standing mode TEG and is illustrated in Figure 1a(i)–(iii) [12]. The sliding motion of two cylinders leads to triboelectrification due to the friction between the PFA film and the outer cylinder. Subsequently, the PFA film and aluminum outer cylinder have opposite negative and positive charges on their surfaces, respectively, due to their different triboelectric polarities and the amount of transferred opposite charges was saturated. The surface charge density of the outer cylinder was twice the surface charge density of the PFA film according to the law of charge conservation, because the contact area of the outer cylinder is two times smaller than that of PFA film. Under the short-circuit conditions, when the outer cylinder was placed on electrode A, the electrical potential difference between the two electrodes was positively maximized (i). When the outer cylinder slid and was placed between electrode A and B, the current flowed from electrode A to electrode B to compensate for the electrostatic induction and made the electrical potential of electrode A equal to that of electrode B (ii). Subsequently, when the outer cylinder reaches electrode B, the electrical potential difference between the two electrodes would be negatively maximized (iii). These stages would be repeated by reciprocating motion of cylinders. PFA film is known to have, not only the most negativity in triboelectric series (so that the largest amount of charges is generated by triboelectrification), but also abrasion resistance and mechanical durability [13–15]. The novel triboelectric and mechanical properties of PFA allow for STEG to have a high output performance with a very long lifetime [16,17]. During reciprocating movement of the outer cylinder with a displacement of 50 mm and speed of 100 mm/s, STEG generated a maximum voltage and current of 100 V and 0.014 µA/cm^2, respectively, when the damping force was 0.6 kgf and the width of electrode A and B were 3 mm (Figure 1b,c). Since STEG had a cylindrical shape and a frictional force against mechanical movement, it could take a role of a damper for the applied force as well as energy harvester. This point is very important because it means that STEG can harvest wasted energy without unnecessary energy consumption.

In order to optimize the output performance of STEG, we investigated the dependence of the output performance on geometrical parameters, such as electrode width. We fabricated several STEGs with different widths of copper electrodes (from 1 mm to 7 mm), but the same gap distance of 1 mm between adjacent electrodes. The output voltage and current were measured by sliding the outer cylinder of each STEG at 200 mm/s, with a damping force of 0.6 kgf and displacement of 50 mm, as shown in Figure 2a. Output voltage and current increased as the electrode width increases from 1 mm to 3 mm, showing a maximum value of 131.9 V and 0.060 µA/cm^2 in root mean square (RMS) value, respectively. However, output voltage and current decreased as the electrode width increased over 3 mm.

The dependence of the output performance on the electrode width in the experimental results was verified using numerical simulations with the same geometric modeling as the experiment. Each electrode, of which the electrode widths were 1, 3, 5, and 7 mm was periodically arrayed and terminated as alternating electrode A and B. The gap distance between neighboring electrodes was 1 mm, and between the moving object and bottom electrode was 25 µm. Experimentally, there was PFA film between the top and bottom electrodes, and triboelectric charges existed on the top surface of the PFA film; however, for simplification, PFA film was not considered and the triboelectric charges (25 µC/m^2) existed on the bottom surface of the moving object. For calculating the electric voltage between electrodes A and B, while the top object was moving at 50 mm/s, the electrical circuit module was coupled. The resistance of the electrical circuit module was set to 100 GΩ for open-circuit conditions. RMS voltage values of devices with different electrode widths are compared in Figure 2b. Corresponding to the experimental results, simulation results showed the highest voltage value when the electrode width was 3 mm. With a fixed gap distance between the electrodes (1 mm), the electrodes were more separated from neighboring electrodes as the width decreased. In the case of an electric field from the charged plane, the electric voltage was proportional to distance; thus, electric voltage increased as electrode width decreased. However, the electrode has finite area, so the edge effect should be considered. As the electrode width decreased, the edge effect became larger compared with the electric field from the charged plane; thus, the electric field became smaller [18,19]. Considering these two effects, 3 mm was the optimum width of the electrode for the highest output performance. The experimental raw data and simulation results related to structure-dependent electric field are illustrated in Figure S2.

Theoretical study on the electric fields in the different geometries offer us an understanding of the reason why STEG could have high output characteristics. To compare the intensity of electric fields and the potential difference versus geometry of the free-standing mode TEG, a COMSOL simulation was conducted on the planar and cylindrical models with the same area and charge density (Figure S3). When the planar model has the geometry infinity plane, the edge effect could be ignored. However, the edge effect is not negligible as there is a non-continuous adjacent charge at the edge of the plane [20]. This characteristic of the planar model induced decreases in the electric field and potential difference (Figure S3a). On the other hand, in the case of a cylindrical model, the edge effect was negligible because the electric field was distributed radially so that they had higher potential differences with the same area and charge density (Figure S3b). A cross-sectional profile of the electrical potential along the surface is compared in Figure S3c. The cylindrical charged surface exhibits larger potential than the planar charged surface and the difference is 100 V or more.

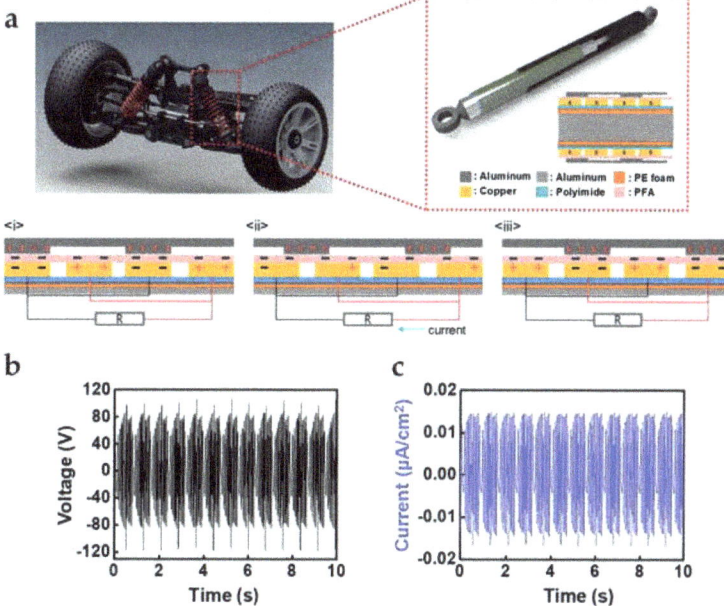

Figure 1. (**a**) Schematic structure and triboelectric regenerative suspension application and the working mechanism of a suspension-type free-standing mode triboelectric generator (STEG); (**b**) the output voltage and (**c**) short-circuit current according to the reciprocating movement of the outer cylinder with a displacement of 50 mm and speed of 100 mm/s.

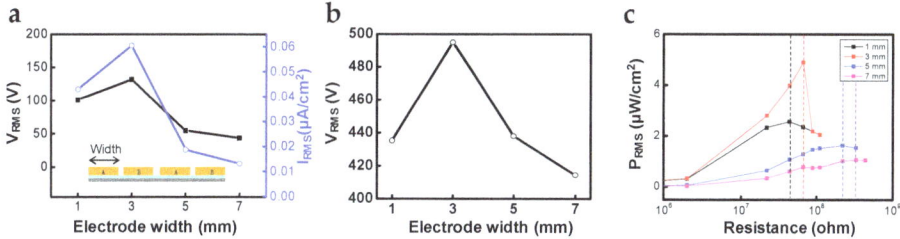

Figure 2. Output performance evaluation and simulation versus electrode width: (**a**) output voltage and short-circuit current; (**b**) COMSOL simulation result (open-circuit voltage) with a resistance of 100 GΩ for open-circuit condition; (**c**) output power in root mean square (RMS) value and different optimum resistance according to changes in the electrode width.

In addition, we determined that each STEG with electrode widths of 1, 3, 5, and 7 mm had different optimum resistances of 44 MΩ, 66 MΩ, 220 MΩ, and 330 MΩ, respectively, as depicted in Figure 2c. As a result, the STEG with an electrode width of 3 mm could generate 4.90 µW/cm² at 66 MΩ. Electrical impedance of the capacitor was inversely proportional to the frequency and capacitance [12,21]. As the electrode width increased, it takes more time for a contacting part of the outer cylinder to move from electrode A to B; as such, the frequency decreased at the same speed when the effective gap between neighboring electrodes increased. Therefore, the optimum electric impedance of STEG increased as a function of electrode width.

3.2. Evaluation of Output Performance depending on Mechanical Input Parameters

Faster speeds of reciprocating movements induced higher output performances because of the short-period of electrostatic induction. The output measurements according to changes in the speed were conducted for STEG, which has a damping force of 0.6 kgf when the displacement is 50 mm and the electrode width is 3 mm (Figure 3a,b). As a result, we found that the maximum short-circuit current and output voltage of 0.060 µA/cm^2 and 131.9 V, respectively, in RMS values could be acquired under a speed of 200 mm/s (Figure 3c). According to reference, which dealt with energy regenerative suspension, the RMS speed of the shock absorber and the speed of a vehicle have a proportional relationship. Through the conversion operation from the speed of the shock absorber to that of a vehicle, the expected short-circuit current in average city driving (32.2 km/h) is about 0.053 µA/cm^2 in RMS value. Although the frictional force is one of the notable factors that has a great influence on the output performance of TEGs, experimental and theoretical studies have not been reported in detail. There have been several reports dealing with changes of output performance in contact mode TEGs depending on increase of applied pushing force [8,22], but the effect of frictional force on sliding mode TEGs has not been investigated because the structural design needed to control and maintain normal force is complex and too strong a frictional force can cause mechanical damage or a reduction in output power due to the disturbance of the movement. However, the cylindrical structure of STEG enables easy control of the frictional force by attaching a polymer elastomer, such as PE foam tape, between the outer surface of the inner cylinder and the flexible PCB layer, as described in Figure 4a. Without PE foam tape, the diameters of the inner and outer cylinders are matched (inner cylinder diameter is slightly smaller than that of outer cylinder), so additional PE foam tape increases the normal and frictional forces according to its thickness. The α is supposed to be the thickness of additional PE foam, and the frictional force increased to 0.5, 0.6, and 0.9 kgf as α was adjusted to 0.08, 0.18, and 0.38 mm, respectively. The output voltage, short-circuit current and frictional force were measured while sliding outer cylinder at a speed of 140 mm/s with a displacement of 50 mm. The output voltage and short-circuit current (RMS value) as a function of frictional force is shown in Figure 4b, and the maximum output voltage and short-circuit current were 137.9 V and 0.057 µA/cm^2, respectively (RMS), with saturating behavior. Experimental raw data related to this issue are illustrated in Figure S4.

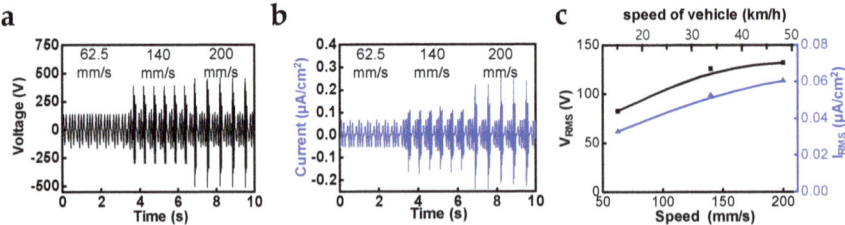

Figure 3. Output performance evaluation versus the different speeds of outer cylinder: (**a**) output voltage and (**b**) short-circuit current with speeds of 62.5, 140, 200 mm/s, respectively; (**c**) output voltage and short-circuit current in RMS value versus the speed of outer cylinder and the speed of a vehicle(expected from the speed of outer cylinder).

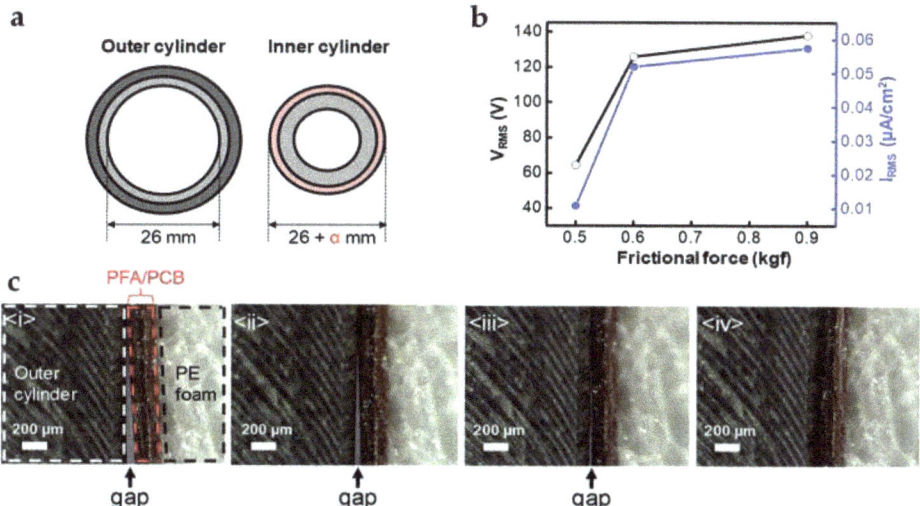

Figure 4. Frictional force as a parameter for output performance of STEG: (**a**) schematic experimental setup to control frictional force through radius control of inner cylinder; (**b**) short-circuit current in RMS value according to increase of frictional force; (**c**) observation by optical microscopy of shape-adapted deformation of PFA film with different α, (i) 0.08 mm, (ii) 0.09 mm, (iii) 0.10 mm, (iv) 0.11 mm, respectively. For α of 0.11 mm, there is no gap.

Due to the elasticity of PE foam tape, it pushes upper flexible PCB and PFA layers to the outer cylinder, which increases the contact area between them, as well as the frictional force. Figure 4c shows a cross-sectional image of the interface between the outer cylinder and the PE foam tape covered by PFA and flexible PCB. The contact area became larger as the thickness of the PE foam tape increased. Triboelectric charges are generated by contact and sliding between the two surfaces. In other words, a larger contact area can generate more triboelectric charges and then output power can be enhanced. However, because the increase in effective area by the applied normal force has certain limitation, the output power vs. frictional force curve has a saturating behavior (Figure 4b). Moreover, too large normal force and frictional force can induce the tearing of the PFA film or disrupt the reliable operation of the suspension. Therefore, we have to choose an optimum frictional force to achieve both high performance and sustainable operation. If the STEG is applied to the suspension system of the vehicle, we can expect much higher output performance because of the damping force of tons' scale and larger size of shock absorber. Considering these expectations, we can observe the feasibility of STEG as a practical energy source for sensors in ADAS technology. Further study should be conducted considering the decay of output performances under high temperature conditions and when facing abrasion problems.

Based on the results of evaluations, a durability test was conducted and we succeeded in charging the capacitors (Figure 5a,b). STEG maintained a stable output performance for 11,000 cycles, even with a damping force of 0.6 kgf, speed of 140 mm/s and a displacement of 50 mm. Inset figures are the short-circuit current peaks at the beginning (red box) and end of the durability test. During the test, the amplitude of the current peaks was maintained at 9.2–9.5 µA. (Figure 5a) Signs of abrasion on the PFA surfaces were not observed because of its abrasion resistance (Figure 5c,d). We also succeeded in charging a capacitor of 330 µF capacitance up to 3 V in 6 minutes at a speed of 200 mm/s, a damping force of 0.6 kgf, and a displacement of 50 mm to charge the capacitor. (Figure 5b). These results mean that the stable and high output characteristics of STEG, not only do not reduce the reliability of existing vehicle suspensions that require a long life, but also show the possibility of efficiently harvesting mechanical energy, which is wasted by the vibration of a vehicle.

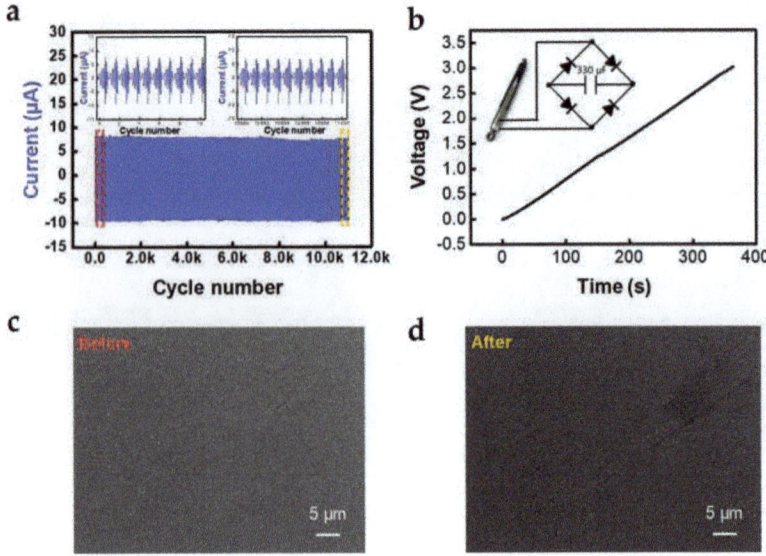

Figure 5. Stability of output performance and charging performance: (**a**) durability test for 11,000 cycles with damping force of 0.6 kgf, speed of 140 mm/s and displacement of 50 mm (inset: short-circuit current signals at the beginning and the end of durability test, respectively; red box is the beginning part and orange box is the end part); (**b**) charging curve for 330 μF using a rectifying bridge, with a frictional force of 0.6 kgf and speed of 200 mm/s; (**c**,**d**) SEM images of surface morphology of PFA film before and after the durability test, respectively.

4. Conclusions

Through an optimization process and evaluations, we could observe the feasibility of STEG to be applied to shock absorber in the suspension systems of vehicles and become an energy source for low-power sensors for ADAS technology according to their stable and high-output performance. Decisively, a novel, suspension-type, free-standing mode structure and the selection of materials considering both industrial and experimental issues, support the impressive reliability of STEG. Proposing the frictional force as a meaningful parameter related to output performance through an experimental approach, we could determine how frictional force could be adjusted, considering not only output performance, but also the stability of the device. Meanwhile, STEG, which was used in this work, was designed to have 1:8 scaled size compared with the actual size of suspension in a commercial vehicle. Additionally, a vehicle applies tons of damping force to the suspension in practical conditions. Considering the much larger scale and damping force applied to the suspension systems in vehicles, it is expected that a much larger output power can be achieved than the experimental results obtained in the present work.

Supplementary Materials: The following are available online at http://www.mdpi.com/2072-666X/10/1/17/s1, Figure S1: detailed structure and optical images of STEG, Figure S2: output performance and COMSOL simulation results versus electrode width, Figure S3: Advantages of cylindrical over planar structure with same charge density and area, Figure S4: output performance versus frictional force.

Author Contributions: Conceptualization, M.K., W.S. and S.-W.K.; Validation, M.K. and W.S.; Formal analysis, M.K. and T.Y.K.; Simulation, T.Y.K.; Methodology support, J.-H.H.; Supervision, S.-W.K.; Writing—original draft, M.K., T.Y.K. and S.-W.K.

Funding: This research was financially supported by a project No SI1802 (Development of One-patch device for HMI based on 3D Device Printing) of the Korea Research Institute of Chemical Technology (KRICT), the GRRC program of Gyeonggi province (GRRC Sungkyunkwan 2017-B05), and Korea Electric Power Corporation (Grant number: R18XA02).

Acknowledgments: Minki Kang and Tae Yun Kim equally contributed to this work.

Conflicts of Interest: The authors declare no conflict of interest.

References

1. Raj, A.; Steingart, D. Power Sources for the Internet of Things. *J. Electrochem. Soc.* **2018**, *165*, B3130–B3136. [CrossRef]
2. Yang, D.G.; Jiang, K. Intelligent and connected vehicles: Current status and future perspectives. *Sci. China Technol. Sci.* **2018**, *61*, 1446–1471. [CrossRef]
3. Xie, X.D.; Wang, Q. Energy harvesting from a vehicle suspension system. *Energy* **2015**, *86*, 385–392. [CrossRef]
4. Zuo, L.; Scully, B. Design and characterization of an electromagnetic energy harvester for vehicle suspensions. *Smart Mater. Struct.* **2010**, *19*, 045003. [CrossRef]
5. Wang, J.; Wu, C.S. Achieving ultrahigh triboelectric charge density for efficient energy harvesting. *Nat. Commun.* **2017**, *8*, 88. [CrossRef] [PubMed]
6. Yoon, H.J.; Ryu, H.; Kim, S.W. Sustainable powering triboelectric nanogenerators: Approaches and the path towards efficient use. *Nano Energy* **2018**, *51*, 270–285. [CrossRef]
7. Kwak, S.S.; Kim, H. Fully Stretchable Textile Triboelectric Nanogenerator with Knitted Fabric Structures. *ACS Nano* **2017**, *11*, 10733–10741. [CrossRef]
8. Seung, W.; Yoon, H.J. Boosting Power-Generating Performance of Triboelectric Nanogenerators via Artificial Control of Ferroelectric Polarization and Dielectric Properties. *Adv. Energy Mater.* **2017**, *7*, 1600988. [CrossRef]
9. Zhang, H.L.; Yang, Y. Single-Electrode-Based Rotating Triboelectric Nanogenerator for Harvesting Energy from Tires. *ACS Nano* **2014**, *8*, 680–689. [CrossRef]
10. Bae, J.; Lee, J. Flutter-driven triboelectrification for harvesting wind energy. *Nat. Commun.* **2014**, *5*, 4929. [CrossRef]
11. Chen, J.; Wang, Z.L. Reviving Vibration Energy Harvesting and Self-Powered Sensing by a Triboelectric Nanogenerator. *Joule* **2017**, *1*, 480–521. [CrossRef]
12. Niu, S.; Liu, Y. Theory of freestanding triboelectric-layer-based nanogenerators. *Nano Energy* **2015**, *12*, 760–774. [CrossRef]
13. Zhang, H.L.; Yang, Y. Triboelectric Nanogenerator for Harvesting Vibration Energy in Full Space and as Self-Powered Acceleration Sensor. *Adv. Funct. Mater.* **2014**, *24*, 1401–1407. [CrossRef]
14. Sidebottom, M.A.; Pitenis, A.A.; Junk, C.P.; Kasprzak, D.J.; Blackman, G.S.; Burch, H.E.; Harris, K.L.; Sawyer, W.G.; Krick, B.A. Ultralow wear Perfluoroalkoxy (PFA) and alumina composites. *Wear* **2016**, *362*, 179–185. [CrossRef]
15. Su, Y.J.; Yang, Y. Fully Enclosed Cylindrical Single-Electrode-Based Triboelectric Nanogenerator. *ACS Appl. Mater. Interfaces* **2014**, *6*, 553–559. [CrossRef] [PubMed]
16. Lee, J.W.; Ye, B.U.; Baik, J.M. Research Update: Recent progress in the development of effective dielectrics for high-output triboelectric nanogenerator. *APL Mater.* **2017**, *5*, 073802. [CrossRef]
17. Kang, H.; Kim, H. Mechanically Robust Silver Nanowires Network for Triboelectric Nanogenerators. *Adv. Funct. Mater.* **2016**, *26*, 7717–7724. [CrossRef]
18. Gallagher, E.; Moussa, W. A study of the Effect of the Fringe Fiels on the Electrostatic Force in Vertical Comb Drives. *Sensors* **2014**, *14*, 20149–20164. [CrossRef]
19. Sloggett, G.; Barton, N.; Spencer, S. Fringing fields in disc capacitors. *J. Phys. A* **1986**, *19*, 2725. [CrossRef]
20. Elrashidi, A.; Elleithy, K.; Bajwa, H. The fringing field and resonance frequency of cylindrical microstrip printed antenna as a function of curvature. *IJWCN* **2011**. Available online: http://citeseerx.ist.psu.edu/viewdoc/summary?doi=10.1.1.678.8736 (accessed on 05 June 2014).
21. Niu, S.; Wang, Z.L. Theoretical systems of triboelectric nanogenerators. *Nano Energy* **2015**, *14*, 161–192. [CrossRef]
22. Lee, J. Shape memory polymer-based self-healing triboelectric nanogenerator. *Energy Environ. Sci.* **2015**, *8*, 3605. [CrossRef]

© 2018 by the authors. Licensee MDPI, Basel, Switzerland. This article is an open access article distributed under the terms and conditions of the Creative Commons Attribution (CC BY) license (http://creativecommons.org/licenses/by/4.0/).

MDPI
St. Alban-Anlage 66
4052 Basel
Switzerland
Tel. +41 61 683 77 34
Fax +41 61 302 89 18
www.mdpi.com

Micromachines Editorial Office
E-mail: micromachines@mdpi.com
www.mdpi.com/journal/micromachines